TEXTBOOK ON
SCIENTIFIC BEEKEEPING
AND VALUE ADDITION OF BEE PRODUCTS

by

N. Emmanuel

A.V.D. Dorajee Rao

K. Gopal

L. Naram Naidu

M. Madhavi

T.S.K.K. Kiran Patro

2026

Daya Publishing House®
A Division of

Astral International Pvt. Ltd.
New Delhi – 110 002

ISBN: 978-93-5919-700-5

Publisher's Note:

Every possible effort has been made to ensure that the information contained in this book is accurate at the time of going to press, and the publisher and author cannot accept responsibility for any errors or omissions, however caused. No responsibility for loss or damage occasioned to any person acting, or refraining from action, as a result of the material in this publication can be accepted by the editor, the publisher or the author. The Publisher is not associated with any product or vendor mentioned in the book. The contents of this work are intended to further general scientific research, understanding and discussion only. Readers should consult with a specialist where appropriate.

Every effort has been made to trace the owners of copyright material used in this book, if any. The author and the publisher will be grateful for any omission brought to their notice for acknowledgement in the future editions of the book.

Published by : **Daya Publishing House®**
A Division of
Astral International Pvt. Ltd.
– ISO 9001:2015 Certified Company –
4736/23, Ansari Road, Darya Ganj
New Delhi-110 002
Ph. 011-43549197, 8130496929
E-mail: info@astralint.com
Website: www.astralint.com

Printed at : **Replika Press Pvt. Ltd.**

Dr. K. Gopal
Vice-Chancellor

Dr. Y.S.R. Horticultural University
Venkataramannagudem,
Tadepalligudem - 534101,
West Godavari District, Andhra Pradesh
Ph:08818-284312
Mobile:+91-7382633666
Email:vc@drysrhu.edu.in
Website:www.drysrhu.ap.gov.in

Preface

Beekeeping has a long history in many civilizations and is a traditional form of livelihood in India and worldwide. India now has over 3.5 million bee colonies, with a growing number of beekeepers meeting the demand for bee products. The number of beekeepers and beekeeping enterprises is rapidly increasing in response to the rising demand for and significance of bee-derived products in the marketplace.

Honey bees are valued for their contributions to the provision of high-quality foods such as honey, royal jelly, and pollen as well as other products utilized in the healthcare and various industries including beeswax, propolis, and honey bee venom. The overall market value of honey is approximately $500 million, with projections indicating that it may escalate to $1,130 million by the year 2025. In Andhra Pradesh, roughly 23.0% of the land is forested, which serves as a crucial resource for honey production.

Keeping this in view, the text book entitled **"SCIENTIFIC BEEKEEPING AND VALUE ADDITION OF BEE PRODUCTS** "has been written and the publication deals with basics and applied aspects of beekeeping. The chapters present an accurate and comprehensive review of scientific beekeeping thus making the text an excellent coverage of practical guide to beginners in apiculture.

Definitely the Textbook on "SCIENTIFIC **BEE KEEPING AND VALUE ADDITION OF BEE PRODUCTS"** is expected to meet the growing needs of undergraduate and post graduate students, KVK experts and beekeepers in the state and gain more impetus for development and up-scaling value added products of honey and beeswax.

I accolade the authors for bringing out the Text book on **"SCIENTIFIC BEEKEEPING AND VALUE ADDITION OF BEE PRODUCTS"**.

(Dr. K. Gopal)

ABOUT THE AUTHORS

Dr. N. Emmanuel is a renowned Professor of Entomology at Dr. YSR Horticultural University, specializing in biological control, integrated pest management (IPM), toxicology, and the innovative use of drones and Artificial Intelligence (AI) for pest management. With over 80 research papers and textbooks to his name, he has earned prestigious accolades such as the Dr. M. S. Venugopal Gold Medal and the Dr. Guruprasad Pradhan Gold Medal. In 2023, he was honored with the Best Teacher Award by Dr. YSR Horticultural University. Dr. Emmanuel has led several impactful projects in coconut, cocoa, beekeeping, and vegetable crop IPM, significantly benefiting farmers. His commitment to advancing scientific beekeeping earned him the Excellence in Innovation Award in 2020 and the Scientist Award in 2021. Dr. N. Emmanuel composed and published Telugu video song "TENETEEGALU" in YouTube to raise awareness about the vital role of honey bees in horticulture, highlighting their importance in pollination and agricultural productivity.

Dr. A.V.D. Dorajee Rao, Associate Dean of Dr. YSRHU College of Horticulture, Venkataramannagudem, is a renowned academic and researcher with gold medals in B.Sc. (Ag.), M.Sc. (Ag.), and Ph.D. degrees. He has vast experience in teaching, research, and extension, publishing over 125 research articles and guiding numerous postgraduate and doctoral scholars. He was instrumental in establishing KVK Nellore and contributed significantly to national coconut research projects. With a proven track record in administration, he has held roles such as Warden, Officer-in-Charge for academics, and Campus Beautification Chairman. As a member of the Board of Studies, he continues to drive academic excellence, research innovation, and campus development at Dr. YSRHU.

Dr. K. Gopal, Vice-Chancellor of Dr. YSR Horticultural University, is a distinguished academician and researcher with 35 years of experience in teaching, research, and administration. He has held key roles such as Registrar, Controller of Examinations, and Zonal Research Head. A pioneer in crop protection, he developed groundnut varieties KRG 2 and KRG 3, disease-tolerant sesame variety Sweta Til, and canker-resistant acid lime varieties "Balaji" and "Vakula." Dr. Gopal established a Molecular Diagnosis Laboratory for virus-free citrus certification and developed Integrated Disease Management technologies. With 143 research publications and 124 conference papers, he has guided 6 Ph.D. and 27 PG students. His numerous awards include the Dr. Shyam Singh Best Scientist Award and Sardar Vallabhbhai Patel International Award. A Fellow of several scientific societies, his work has advanced agricultural and horticultural sciences in India.

Dr. L. Naram Naidu, Dean of Horticulture at Dr. YSR Horticultural University, A.P., is an alumnus of the Agricultural College, Bapatla under Acharya N.G. Ranga Agricultural University, Andhra Pradesh. With over 35 years in research, teaching, and extension, he has served in various academic and administrative positions, including Zonal Research Head, Director of Research, and Dean PG Studies. Developed seven chilli varieties, and contributed to improved spice crops. He played a key role in establishing research stations, securing external funding, and developing infrastructural facilities. Dr. Naidu has co-authored two textbooks, published 120 research papers, and presented 115 conference papers. Recognized for his contributions, he received the Best Seed Scientist award from the Seedsmen Association, Hyderabad, for his impact on the seed industry and farmers' service.

Dr. M. Madhavi earned her B.Sc. (Hort.) in 1987, M.Sc. (Hort.) in 1989, and Ph.D. (Hort.) in 2007. She is presently working as Director of Research at Dr YSR Horticultural University, venkatramannagudem, Andhra Pradesh. With over 10 years of research experience in citrus crop cultivation and 20 years in teaching, Dr. M. Madhavi is a renowned expert in the field of horticulture. She has guided 25 M.Sc. and 12 Ph.D. students and has published over 55 research papers, 20 popular articles, and 6 book chapters. She is also a member of various horticultural societies, showcasing her commitment to the field.

Dr. T.S.K.K. Kiran Patro is an Associate Professor of Horticulture and Technical Officer to the Dean at Dr. YSR Horticultural University, Venkataramannagudem. She holds a BSc in Agriculture, an MSc (Ag) in Horticulture, and a PhD from ANGRAU. With expertise in vegetable crop research, she has guided several M.Sc. students on crops like pointed gourd, gherkins, and potato. She has led TSP and SCSP projects focused on cashew, IPM for vegetables, and beekeeping to support farmers. She has published over 50 research papers and contributed to knowledge dissemination. In recognition of her work, she received the Young Scientist Award for significant advancements in vegetable crop sciences.

Contents

CHAPTER 1

History and Development of Beekeeping

Bees have much to teach us, they enthrall us, they are beautiful, and their significance to mankind is beyond measure. Undeniably it is impossible to imagine life without sweetness, especially since craving sweet sensations is instinctive for humans.

In order to be able to feed the world's growing population, we need ever more food, which must be diverse, balanced and of good quality to ensure the progress and well-being of humankind. Bees are renowned for their role in providing high-quality food (honey, royal jelly and pollen) and other products used in healthcare and other sectors (beeswax, propolis, honey bee venom). But the work of bees entails much more! The greatest contribution of bees and other pollinators is the pollination of nearly three quarters of the plants that produce 90% of the world's food. A third of the world's food production depends on bees, i.e. every third spoonful of food depends on pollination.

In traditional medicine, people used honey when treating a wide variety of conditions. While many of these uses do not have scientific backing, they include: eye diseases, bronchial asthma, throat infections, tuberculosis, thirst , hiccups, fatigue, dizziness, hepatitis , constipation, worm infestation, hemorrhoids, eczema, ulcers, wounds. Beeswax is another important product that people have previously used in waterproofing and fuel. It currently has benefits for health and features in a number of skincare products. Additionally, Pharmaceutical industries used it in ointments. Other bee products that can benefit human health include: propolis, a resin that bees create bee bread, bee pollen, royal jelly, beeswax, bee venom. Scientists found evidence that melittin, a component in honeybee venom, could kill cancer cells.

Beekeeping is the maintenance of honey bee colonies in man-made hives. A location where bees are kept is called an apiary or "bee yard". The majority of the honey bees are included in the genus *Apis*, but other honey-producing bees for instance *Melipona* stingless bees are also maintained. A beekeeper keeps bees to collect their honey and other by-products such as beeswax, propolis, bee pollen, and royal jelly. Of late, beekeepers are renting their colonies for the pollination of Horticultural and Agricultural crops as well.

Humans likely gathered honey in the same manner as their primate ancestors, but the first representations of interaction between humans and bees are in cave paintings of Spain, India, and Africa that date to approximately 8,000 BCE. The paintings show human figures gathering wild honey from nests hanging on trees but not from beehives.

After the discovery of this marvelous substance came efforts to "cultivate" it, one of the most significant developments in the agro-pastoral way of life: housing bees inside pots, namely beekeeping. According to iconographic evidence, this important innovation took place in Egypt somewhere around 2,500 BCE. Representations in tombs of officials show cylindrical, horizontally placed clay pots containing hives, and illustrate the collection and storage of honey. Ancient Egyptians developed beekeeping as a large-scale production process and expanded the uses of bee products. Honey was not only used in cooking, baking, and beverage making, but in medicine and pharmacology, including treatment of wounds and various illnesses, in mummification, and in beauty and cosmetic products, such as perfumes. One of the most famous examples is Cleopatra's bath in milk and honey.

Beekeeping then diffused to Mesopotamia, Europe, China, and India. The excavations in northern Israel at Tel Rehov, which date to the mid-tenth century BCE, uncovered the remains of a huge apiary, consisting of at least 180 horizontal, cylindrical beehives made of sun-dried mud. This discovery provides a tangible link to the Old Testament phrase to the Promised Land flowing with milk and honey. Honey is also mentioned in other religious texts, such as the Vedas, the Talmud, the New Testament, and the Quran. For example, Hindu gods Vishnu, Krishna, and Indra were called *Madhava*, nectar-born ones and were symbolized by bees. In the Quran, Surah is entitled "The Bees" and commends humans to emulate the industry of bees. Beyond its nutritional value, honey became a spiritual and metaphysical symbol, which affected the collective consciousness of many traditions.

Traces of beeswax on prehistoric potsherds have revealed that the harvesting of bee products by man has been practiced in Greece since the Middle Neolithic period. In ancient Greece, the most famous production centre of thyme honey – the finest variety – was Athens. Ancient writers even attributed supernatural qualities to Attic honey, believing it could bring long life. Philosophers such as Pythagoras and Democritus – both of whom lived until approximately 90 – claimed that their good health was due to the consumption of honey. For the first time honey also became the subject of scientific research by the Greek philosopher Aristotle, the first scientific beekeeper, in his compendium *Historia animalium*. The physician Hippocrates also established the health benefits of honey, advising the use of vinegar and honey for pain, water and honey for thirst, and a mixture of honey, water, and other substances for fevers. Hippocrates recommended honey for treatment of many other conditions, including baldness, contraception, wounds, eye diseases, and many more. Both honey and the bees played a central role in ancient Greek myths, as symbols of chastity, rebirth, and fertility.

The Romans dealt with beekeeping extensively, organizing it in terms of taxation and achieving high levels of quality, variety, and efficiency. Bees and the beekeeping practices were thoroughly studied by Roman agricultural writers such as Virgil, Varro, Columella, and Pliny the Elder. The physician Pedanius Dioscorides described the characteristics and qualities of the finest honey as sweet and pungent in taste, with a pleasant smell, yellowish in colour, not liquid but somehow 'sticky' and fine.

Recipes featuring honey are found in the collection *De Re Coquinaria*, erroneously attributed to Marcus Gavius Apicius. These include honey as a substance for the preparation of beverages, as a meat- and fruit preservative, and as an additive to sauces, usually next to vinegar. Souffles, fruit desserts, pancakes, veal, and tuna dishes all featured honey.

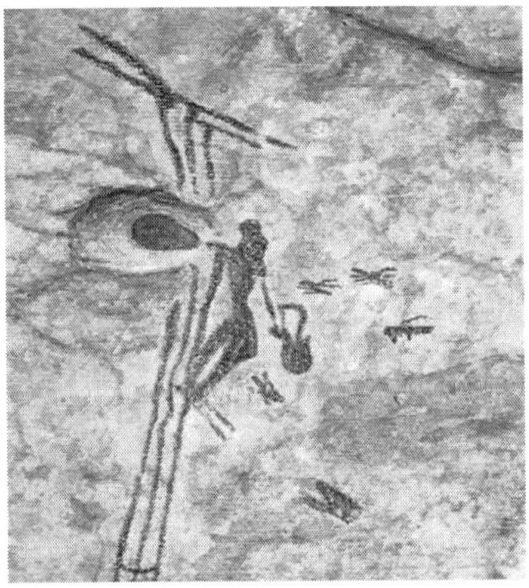

Prehistoric rock-painting of honey gathering, Arana cave, Spain, 8,000 BCE

Beekeepers of Ancient Egypt

Prehistoric bas-relief of a beekeeper harvesting honey, from Egyptian tomb, Pabasa, 7th-6th century BCE.

26th Egyptian Dynasty hieroglyph depicting a beekeeper with their hives of bees

By the end of Classical Antiquity and during the spread of Christianity, honey collection and management were fully organised industries. Mediterranean honey producers sought ways to increase the quantity to be gathered and to improve the honey's quality. In order to attain these goals, simple but effective practices were applied, such as nomadic beekeeping, in which beekeepers moves hives from place to place to follow the various seasonal plants in bloom. Another method to improve quality were technical innovations, probably applied for the first time by the beekeepers in ancient Attica, namely vertical rather than horizontal beehives, and circular extension rings adjusted the mouth of horizontal hives.

Traditional Hives Around the World Include

Mud hives: Mud hives are still used in Egypt and Siberia. These are long cylinders made from a mixture of unbaked mud, straw, and dung.

Clay hives: Clay tiles were the customary homes of kept bees in the eastern end of the Mediterranean. Long cylinders of baked clay were used in ancient Egypt, the Middle East and to some extent in Greece, Italy and Malta. They sometimes were used singly, but more often stacked in rows to provide some shade, at least for those not on top. Keepers would smoke one end to drive the bees to the other end while they harvested honey.

Skeps : Skeps, baskets placed open-end-down, have been used to house bees for some 2000 years. Believed to have been first used in Ireland, they were initially made from wicker plastered with mud and dung but after the Middle Ages, almost all were made of straw. In northern and western Europe, skeps were made of coils of grass or straw. In its simplest form, there is a single entrance at the bottom of the skep. Again, there is no internal structure provided for the bees and the colony must produce its own honeycomb, which is attached to the inside of the skep.

Traditional Hives of India

Clay Pots : Bees are reared in clay pots in parts of south India. Air holes are punched in the pots, smeared with beeswax and kept in gardens. They act as instruments to attract the swarms. Once the colony is set, another pot is inverted over this. Once the rains set in the pots are overturned and honey is harvested.

Tree Trunks : Trunks of trees or hollow wooden logs are used as hives. However, this is in altitudes above 1800m sea level.

Wall Hives : These are common in the northern states like Jammu and Kashmir, parts of Himalayas etc. They are basically empty spaces on the eastern side wall of the house.

Bee coin - Ephesus used the bee on its coins since it was a producer of honey

Ancient cylindrical beehives made of sun-dried mud from Israel

Wall painting of the tomb of Rekhmire (c. 1450 BCE) in Egypt, depicting horizontal beehives

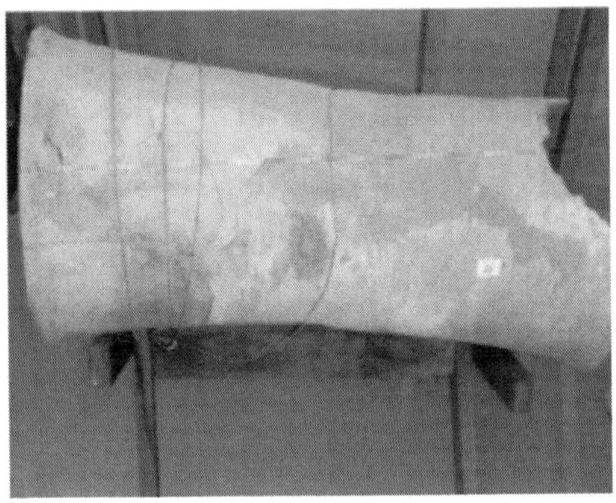

A traditional horizontal beehive from Greece (photo H. Harissis, Collection of A. Bikos in the Geoponic Institute, Athens)

It was not until the 18th century that European understanding of the colonies and biology of bees allowed the construction of the movable comb hive so that honey could be harvested without destroying the entire colony. Johannes Mehring of Germany, who invented a press to make wax sheets having the indentations of the bottoms of the cell. This led ultimately to the perfecting commercial methods of making comb foundation. The third man to be considered was the Austrian Franz von Hruschka, who is credited with the invention of the honey slinger in 1865. This discovery led to the manufacture of the honey extractor, both in Europe and in the United States. During nineteen century, the contributions of American investigators were of immense. In 1851, Reverend Lorenzo Langstroth designed the movable frame bee hive and considered him to be the father of apiculture,

Moses Quinby, the father of commercial apiculture; C. C. Miller, the Nestor of American bee keeping; and A. I. Root, the first commercial manufacturer of bee equipments in the world.

The Beekeepers, 1568,
Painting by Pieter Bruegel the Elder

Prehistoric smoking pots from greece

Stone bee hive at Malta, ancient Roman City

Part of the prehistoric apiary excavated at Tel-Rehov valley, Israel, 10th-9th century BCE

Bee house in Slovenia

Bee hives in Trees , Ethiopia

Bamboo hive

Banana fiber hive

Straw bee hive

Skep

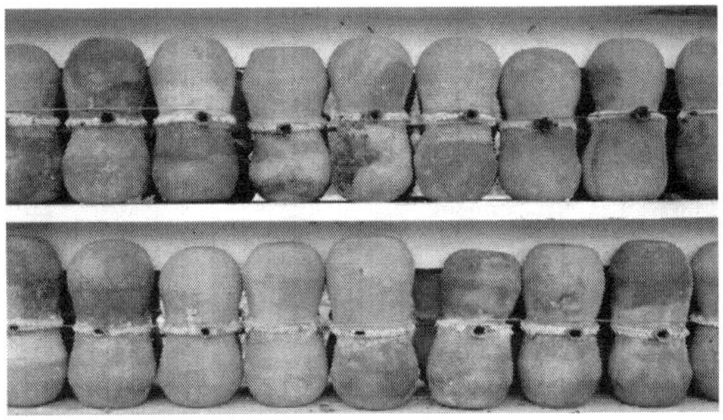

Pot hive with Stingless bees

Barrel Hive with Circular frames

Wooden bee hive with straw

Pot hive

Log hive

Log hive with legs

Vertical log hives

Rev. Lorenzo Lorraine Langstroth was an American apiarist, clergyman, and teacher, and considered to be the father of American beekeeping. He created the modern day Langstroth hive. The Leaf Hive, invented in Switzerland by François Huber, was a fully movable frame hive, but had solid frames that were touching and made up the "box." The combs in this hive were examined like pages in a book. Langstroth acknowledged Huber's contribution and noted: "The use of the Huber hive had satisfied me that, with proper precautions, the combs might be removed without enraging the bees, and that these insects were capable of being tamed to a surprising degree. Without knowledge of these facts, I should have regarded a hive permitting the removal of the combs as quite too dangerous for practical use." (*Langstroth on the Honey-Bee,*). Langstroth was popularly credited with discovering the "bee space," though this discovery had already been implemented in European hives. In Europe, both Jan Dzier on and August von Berlepsch had been focused on side-opened hives. Land resources in Europe had been limited, and bees were traditionally kept in beehouses. The presently so-called "bee space" had been incorporated by Berlepsch following Dzierzon's discoveries, from the years 1835–1848, into his frame arrangement. Langstroth made many other discoveries in beekeeping, and contributed greatly to the industrialization of modern beekeeping.

Langstroth revolutionized the beekeeping industry by using bee space in his top-opening hive. In the summer he found that by leaving an even, approximately bee-sized space between the top of the frames holding the honeycomb and the flat cover board above, he was able to remove the coverboard quite easily, which was normally well-cemented to the frames with propolis, and made separation difficult. Langstroth later used this discovery to make the frames themselves easily removable. If a small space was left (less than 1/4 inch or 6.4 mm), the bees filled it with propolis; conversely, when a larger space was left (more than 3/8 inch or 9.5 mm), the bees filled it with comb.

Langstroth received a patent on the first movable frame beehive in America. A Philadelphia cabinetmaker and fellow bee enthusiast, Henry Bourquin, made Langstroth's first hives for him. Langstroth had more than a hundred of these hives, and began selling them where he could. Langstroth spent many years attempting to defend his patent without success. He never earned any royalties, because the patent was easily and widely infringed. Langstroth hives are still in common use today.

The art of beekeeping appeared in ancient China for a long time and is hardly traceable to its origin. In the book, 'Golden Rules of Business Success' written by Fan Li and Tao Zhu Gong during the Spring and Autumn period, there were some parts mentioning the art of beekeeping and the importance of the quality of the wooden box for bee keeping that can affect the quality of its honey. The ancient people of Maya civilization domesticated a separate species of stingless bee. Michener in his book ' The Bees of the World' revealed a diverse fauna that numbers more than 17,000 species and ranges from the common honeybee to rare bees that feed on the pollen of a single type of plant. Moreover new facts, reclassifications, and revisions provide the most comprehensive treatment of the 1,200 genera and subgenera of the Apiformee. Oldroyd and Wongsiri gave a detail account of Asian honey bees with their evolution, speciation, reproductive and nesting habit of different honeybee species. Further, they mentioned the various factors affecting both domesticated and wild honey bee population with their conservation strategy. Bread bear prepared a guide to the services provided by bees and the sustainable harvesting, processing and marketing of their products and included in the publication of the Food and Agriculture Organization as 'Bees and Their role in Forest Livelihoods'.

REV. LORENZO LORRAINE LANGSTROTH
(December 25, 1810 – October 6, 1895)

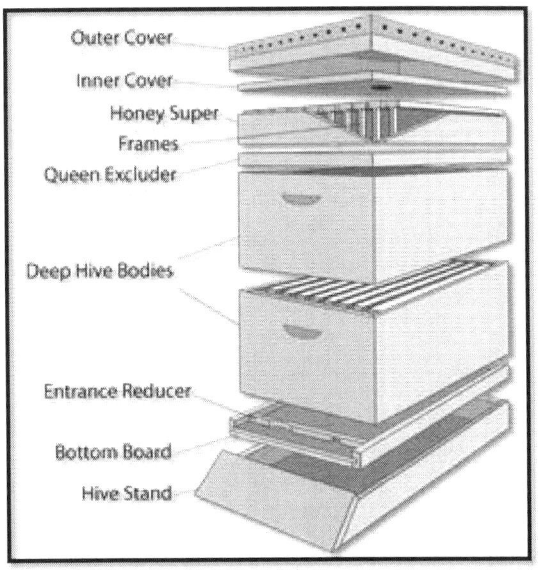

LANGSTROTH BEE HIVE

Pioneers of Practical and Commercial Beekeeping

Beekeeping has seen improvements in the design and production of beehives, systems of management and husbandry, stock improvement by selective breeding, honey extraction and marketing, in the 19th century. Notable innovators of modern beekeeping include:

Petro Prokopovych	He used frames with channels in the side of the woodwork; these were packed side by side in boxes that were stacked one on top of the other. The bees traveled from frame to frame and box to box via the channels. The channels were similar to the cutouts in the sides of modern wooden sections.
Jan Dzier on	Father of modern apiology and apiculture. All modern beehives are descendants of his design.
François Huber	He made significant discoveries regarding the bee life-cycle and communication between bees. Despite being blind, Huber brought to light a large amount of information regarding the queen bee's mating habits and her contact with the rest of the hive. He published as *Observations on the Natural History of Bees.*
L. L. Langstroth	Revered as the «father of American apiculture»; no other individual has influenced modern beekeeping practice more than Lorenzo Lorraine Langstroth. His classic book *The Hive and Honey-bee* was published in 1853.
Moses Quinby	Often termed «the father of commercial beekeeping in the United States,» author of *Mysteries of Bee-Keeping Explained*. He invented the Bee smoker in 1873.
Dr. C.C. Miller	He was one of the first entrepreneurs actually to make a living from apiculture. His book, *Fifty Years Among the Bees*, remains a classic, and his influence on bee management persists to this day.
Franz Hruschka	He was an Austrian/Italian military officer who made one important invention that catalyzed the commercial honey industry. In 1865 he invented the simple machine for extracting honey from the comb by means of centrifugal force. His original idea was to support combs in a metal framework and then spin them around within a container to collect honey as it was thrown out by centrifugal force.
Walter T. Kelley	An American pioneer of modern beekeeping in the early and mid-20th century. He greatly improved upon beekeeping equipment and clothing
Brother Adam	Inventor of the Manley frame, still universally popular in the U.K.
Abushady	Patented a removable, standardized aluminum honeycomb. In 1919 he also founded The Apis Club in Benson, Oxfordshire, and its periodical Bee World, which was to be edited by Annie D. Betts and later by Dr. Eva Crane. The Apis Club was transitioned to the International Bee Research Association (IBRA). Its archives are held in the National Library of Wales. In Egypt in the 1930s, Abushady established The Bee Kingdom League and its organ, The Bee Kingdom.

History of Apiculture in India

Beekeeping in India is as old as Indus valley civilization. The Honey bees were referred in almost all religious texts such as the Vedas, Ramayana, Mahabharata, Quran, Bible etc. The knowledge based on bees, honey and honey harvesting had been in operation through the ages. During the Vedic period, large information was available on different species of honey bees, their habits, different flora that offered nectar and pollen to bees. Honey had been included as apanchamrit in the diet of devatas. With the development of civilization, honey acquired a unique status in the life of the ancient Indians. They considered honey as a heavenly substance that cured many diseases and contributed in vigour and vitality of human beings.

First attempt to keep bees in movable frame hive were made in 1882 in Bengal and in 1883-84 in Punjab but with little success. In 1883, government of India published the information on beekeeping, collected from the provinces, under the title "a collection of papers on beekeeping in India". A year later, John Doughlas, an employee of the Telegraph Department, published A Handbook on Beekeeping in India. The recommendation of the Royal Commission on Agriculture (1928) for developing cottage industries gave a fillip to beekeeping. Work on this subject in right earnest was taken up in Madras in 1931, Punjab in 1933, Coorg in 1934 and Uttar Pradesh in 1938. Other states and provinces followed suit but no important landmark was made. Beekeepers of India organized themselves and formed an All India Beekeepers' Association during 1938-39. Since then the Association has been publishing the Indian Bee Journal.

The Indian Council of Agricultural Research co-ordinates the activities of research work in beekeeping. The Council established a beekeeping research station in the Punjab in 1945 and six years later at Coimbatore in Tamil Nadu. From 1950, the ICAR has been funding various research projects on honey bees in the universities and other organizations. Besides, since 1980 it has started an All India Co-ordinated Research Project on Honeybees Research and Training. During, the past three decades' State Universities, Khadi and Village Industries Commission (KVIC) have also contributed to the development of apiculture. The Central Bee Research and Training Institute was established in 1962 in Pune under aegis of KVIC.

Three *Apis mellifera* colonies were imported from Italy and maintained in a small apiary by Ghosh. However, due to various factors causing queen mating failure, the colonies were lost. Similarly, Thomson documented several unsuccessful attempts to introduce *Apis mellifera* bees in different regions of India. Baldry attempted to establish *Apis mellifera* in Mahabaleshwar, but the colonies were later relocated to Coimbatore, where they eventually perished. In the 1930s, Thomson also brought exotic bees to Kerala, but they failed to thrive, which he attributed to the area's unfavorable climatic conditions for *Apis mellifera* bees. Beadnell brought many consignments of *A. mellifera* and maintained an apiary in Nilgiri Hills. Rehman and Singh received two consignments of two and three colonies and attempted to interbreed with *A. cerena* but it also failed. Both *Apis cerana and Apis mellifera* did not accept each other's brood, workers or queen. *Apis mellifera* colonies were imported to Kashmir in 1951. The queens were introduced to *A. cerana* colonies and were reported to be doing well until 1959. After that there was no report on the progress of these bees, and it is presumed that they have perished. Successful introduction and establishment of *Apis mellifera* was made possible in sixties. *Apis mellifera* and its hybrids were obtained from U.S.A. and Italy between 1962-1964 and were introduced in Nagrota, Himachal Pradesh.

In the line of NCA recommendation, Indian Council of Agricultural Research (ICAR) conceived a project entitled 'All India Coordinated Project on Honeybee Research and Training' in 1980-81. At first, the Project started functioning with the Project Coordinating Unit at Central Bee Research and Training Institute, Pune with six centres including one in an ICAR institute. Five more centres, four State Agricultural Universities (SAU) and one in an ICAR Institute were sanctioned in 1983. Project Coordinating Unit was shifted to Haryana Agricultural University, Hisar in 1987. There were nine University-based coordinating centres during the X Five- Year Plan at Solan (Himachal Pradesh), Ludhiana (Punjab), Vijayarai (Andhra Pradesh), Jorhat (Asom), Pusa-Samastipur (Bihar), Vellayni (Kerela), Hisar (Haryana), Pantnagar (Uttrakhand) and Bhubaneshwar (Odisha).

During the course of time, several apiculture literatures have come up both in popular and scientific series in India. Some of the salient literatures are: Singh contributed a book on Indian beekeeping which included beekeeping practices; management of honeybee species, bee pasturage, disease and enemy management of honey bees in India. Verma wrote a book on Beekeeping in integrated mountain development: economics and scientific perspectives which is a testimony of integrated programme on mountain development in which apiculture was included as component. Later on Mishra gave a comprehensive knowledge on honey bees and their management in India and published under the aegis of Indian Council of Agricultural Research in which he has highlighted research and management activities on different honey bee species. Joshi in her bibliography of Doctoral dessertations on bees, beekeeping and related aspects highlighted different areas of apicultural research such as mellitopalynology, bee botany and pollination, bee management, bee pathology and bee systematics etc. Atwal contributed a book on 'Essentials of Beekeeping and Pollination' which highlighted beekeeping as essential ingredient in agricultural production system.

Dr Rahman contributed the chapter on 'Apiculture' in the Handbook of Agriculture published by the Indian Council of Agricultural Research, New Delhi which was revised in 2009 and 2012, gave a comprehensive account on Indian Apiculture. Dr. Abrol in his book 'Bees and Beekeeping in India' gave an elaborative scenario of both domesticated and wild bees in India. Thakur compiled the research activities on apiculture in India under Apiculture Research and published as Abstract of Research Publication, 2000-2012 under the aegis of Indian Council of Agricultural Research, New Delhi. The publication encompassed recent information on various aspects of apiculture, viz. bee forage and foraging behaviour, pollination, ecology, bee biology and morphmetrices, bee products and technology, bee diseases and enemies, pesticide toxicity, bee keeping equipments etc. Abrol contributed the book on Asiatic honeybee, *Apis cerana:* Biodiversity Conservation and Agricultural Production. A detail Monograph on Morphometry and Phylogeny of Honeybees and Stingless bees of India' was published under the aegis of Department of Biotechnology, Government of India, New Delhi.

CHAPTER 2

Honeybee Species in India

Honey bees are classified in the order Hymenoptera, which also includes ants, bees, sawflies, and wasps. The genus *Apis* consists of eight Asian species and one western species. Ancient bees first emerged 120–130 million years ago (mya) coincident with the emergence of early angiosperms. The ancestors of modern bees were geographically isolated in the Middle East during the late Pleistocene approximately 1–2 million years. It was from this population that the ancestral lineage of *A. mellifera* spread to Africa, and the ancestral lineages of *A. cerana* expanded throughout Europe and Asia. As a result of allopatric speciation, these two honey bee species may have evolved in different ecological environments, which gave rise to different behavioral and physiological characteristics observed in the present day.

Rock Bee (*Apis dorsata*)

They are large and fierce/violent bees that build a single comb in the open typically about 3-4 feet tall. They are around 18–20 mm long. Nests are primarily built in uncovered places far off the ground, on trees, beneath cliff extensions, and occasionally on buildings. These social bees are known for their hostile defense tactics and cruel behavior when bothered. Aboriginal peoples have conventionally used this species as a source of honey and beeswax, traditionally called as honey hunting since prehistoric times.

A. dorsata are found all over the Indian subcontinent mostly in the forests and also in urban areas with reasonable vegetation. In mountainous areas they build their nest up to an altitude of 2700 m. Rock bees routinely shift their nesting places. Annually, 50–80 kg of honey can be harvested from a single comb / colony of *A. dorsata*. In Indian subcontinent, they are found in Pakistan (and, perhaps, parts of southern Afghanistan) in the west and Sri Lanka to Indonesia and parts of the Philippines in the east. Its north-south distribution ranges from the southern part of China to Indonesia; it is not found in New Guinea and Australia.

The following Apis species identified in the Asian region are as follows:

1	*A.d. dorsata* (Indian giant honey bee)	India
2	*A.d. binghami* (Indonesian giant honey bee),	Malaysia and Indonesia
3	*A.d. breviligula* (Philippine giant honey bee)	Philippines
4	A. d. laboriosa (Himalayan giant honey bee)	India, Nepal, Bhutan, Myanmar, Laos, Vietnam and southern China.

As a rule these bees build their combs at a height more than 20 ft above the ground, but in few instances, they are found hanging from branches just above 2 ft from the land. Colonies of rock bee may be found singly or in groups. The underneath part of the comb is the active region where the foraging and scout bees fly off and land. As these rock bees are violent, they will assault the trespasser and every so often they will chase even up to 100 m. seldom rock bee stings can be deadly to the humans. As great jeopardy and risk is involved in harvesting *A.dorsata* honey, it fetches very high price in the local market. Skilled bee hunters choose to harvest at night. Smoke is gusted to calm down the ferocious rock bees, and in a few places locals add fowl feathers to smoke formed by blazing charcoal which infuriates the bees and make them to migrate and furthermore the horrible smell produced due to addition of fowl feathers, allowing effortless honey extraction. There is universal apprehension that the entire number of *A. dorsata* population in all Asian countries is diminishing, to a certain extent due to attenuation of forest expanse, use of noxious pesticides in agriculture and roc bee hunting.

Indian Bee (*Apis cerana*)

Indian Bee (*Apis cerana*) Indian honeybee is a renowned bee species in India. Earlier to the ingress of *A.mellifera* , it was barely domesticated honey bee spp. in India. *A.indica* is also observed in Sri Lanka Nepal, Burma, Bangladesh Thailand and Pakistan. *A.indica* are relatively non-aggressive and hardly ever shift positions. *A.indica* construct multiple parallel combs in dark such as clay pots, logs, wall, tree openings, etc. and generate 7–9 kg of honey per colony per year. *A. cerana* is the sister species of *Apis koschevnikovi* and together are in the identical subgenus as the European honey bee, *Apis mellifera*. *A. cerana* is known to live sympatrically all along with *Apis koschevnikovi* in the similar geographic location.

Indian Bee, *Apis cerana* has numerous subspecies based on genetic diversity and survival environment. Amongst these *Apis cerana cerana* and *Apis cerana indica* are found in India. The subspecies *Apis cerana indica* is documented into two morphotypes like black coloured "hills bee" and yellow coloured "plains bee". At the moment beekeeping by means of Indian bees is regularly done in south India and predominantly in Kerala and Tamil Nadu, with more than one lakh beekeepers occupied with ardent dedication. Given that these bees have constructed their combs in dark cavities, it facilitates beekeepers to domesticate them in specifically build movable frame hives. *A. cerana* colonies are constructed in parallel combs at consistent expanse recognized as the "bee space".

Contrast to rock bees and Italian bees, Indian Bee, *Apis cerana* are small in size but larger than the dwarf bees. Brood comb has cells of two dissimilar sizes: larger for the drone brood and smaller for the worker brood. The queen cells are constructed on the lower perimeter of the honey comb. Analogous to other honey bee species, *Apis cerana* bees also accumulate honey in the upper part of their comb. Owing to this behavior, the bee boxes are designed and fabricated in such a manner that the super chamber or the honey chamber is placed in the upper part of the hive where these honey bees stock up honey which help out in trouble-free honey extraction.

Compared to *A. mellifera*, the Asian honey bee, *A. cerana*, has quite a lot of distinctive behavioral qualities. It is able to acclimatize to intense weather conditions and has long flight length, effectual grooming clean and healthy behaviors with elite cooperative and collective defenses/ resistance. A distinguished behavior of *A. cerana* is aggregation when a colony is exposed to threats, such as predators or intruders. Under these situations, worker bees which are guard bees emit out alarm pheromones that influence group actions.

European Bee/Italian Bee (*Apis mellifera ligustica*)

Italian honey bee, *Apis mellifera ligustica*, is mainly mantained by beekeepers in the western hemisphere of the globe. Despite the fact that inhabitant to Europe and Africa, *Apis mellifea* is now scattered worldwide, principally due to beekeeping. Italian honey bees are illustrated as light or golden in color. They have striped yellow and brown abdomens and heads are hairy with large compound eyes and ringed with hair.

The average size of the worker bees ranges from 10–15 mm long, while queens are larger (18–20 mm), and drones reach 15–17 mm at maturity. *A. mellifera* workers are by and large reddish-brown (based on the race, some are lighter and some are darker) with dark bands in the metasoma; workers legs are dark brown.

Italian bee (*Apis mellifera ligustica*) is not native to India but was introduced from West as the native Indian bee colonies were disappearing due to Thai sacbrood virus. Currently they are successfully established in India and abundantly present in northern India due of the rich flora such as mustard, safflower, sun flower, etc. As rice is the major crop in south India, these bees don't get enough amount of food they need. In south India, beekeeping with Italian bees is difficult to domesticate but now its being practiced under shady plantation crops and in forest ecosystems; for commercial purpose, these bees must be relocated based on floral mapping. They are also comparable in habits to Indian bees, which build parallel combs in dark places and stock up honey at the upper part of their hive. They are bigger than all other honeybees except *Apis dorsata*. Italian bee produce 25–40 kg of honey per colony per year. Perhaps these bees are the one of the most studied bees. The introduction of *A. mellifera* to India created tribulations such as the interspecies transmission of pests and diseases of honey bees. However, introduction of *Apis mellifera* bees into India can be documented as triumphant history because it generated employment for quite a lot of people with cost-effective profits and also by the pollination service these bees done to Indian agriculture.

Beekeepers have preferred italian honey bees for a number of enviable features: capability of a colony to endure extreme conditions with scanty food, the ability of a colony to survive freezing weather, resistance to infections and disease, amplified honey production, reduced ferociousness, reduced tendency to swarm, reduced hive building, effortless pacification with smoke.

These alterations, all along with artificial change of location, have enhanced domestication of western honey bees as per beekeeper perspective, and concurrently made them more reliant on beekeepers for their continued existence. In Europe, chilling weather survival was likely preferred for, deliberately or not, whilst in Africa, selection most likely privileged the capacity to survive high temperature, famine and heavy rainfall.

ROCK BEE – *Apis dorsata*

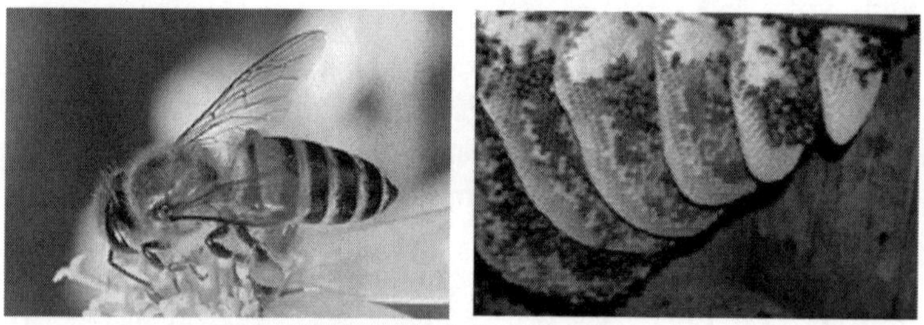

INDIAN BEE – *Apis cerana indica*

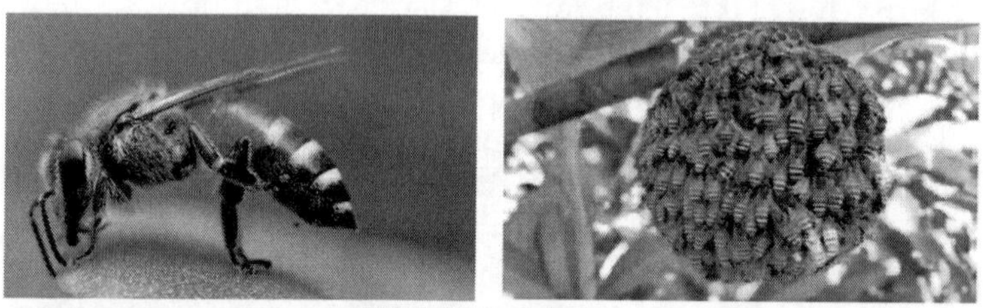

LITTLE BEE – *Apis florea*

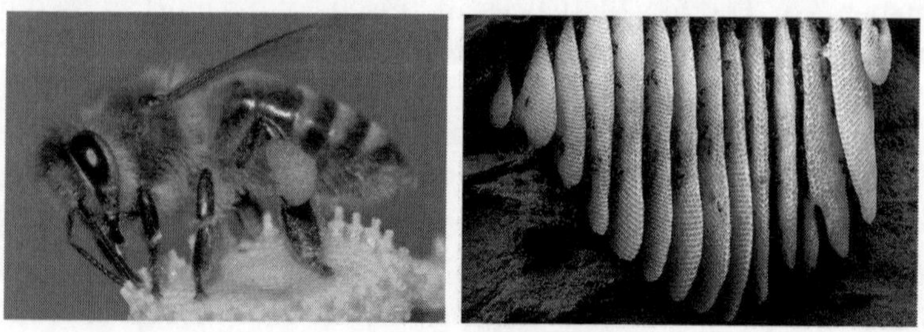

ITALIAN BEE – *Apis mellifera*

Little Bee (*Apis florea*)

Apis florea or dwarf honeybee is also a wild honeybee spp., but these bees are small and less ferocious when compared to the rock bees. *A. florea* is called the dwarf honey bee due to its small size compared to other honeybees. A worker is typically 7–10 mm in body length and its overall coloration is red-brown. Drones carry a thumb-like bifurcation called the basitarsus, which is located two-thirds along the length of the tibia. The fimbriate lobe of *A. florea* has three protrusions and they sting using two stylet barbs.

These bees build single vertical combs they also construct palmsized combs in the bushes, hedges, buildings, caves, empty cases, etc. The major difference between the rock bee and little bee comb is that the little bees construct combs encircling the twigs while the rock bees construct the comb on the undersurface of the branch. The honey produced by these bees is dramatically less when compared to the rock bee as these bees produce only about half a kilo of honey per year per hive. However, in the Kutch area of Gujarat, large quantities of honey from *A. florea* are harvested. As these bees also have a habit of shifting their colonies frequently, they are also non-rearable, but attempts in India have brought partial success.

These bees are found only in plains and not in hills above 450 MSL. Compared to other honeybees, these bees are attractively coloured with red to brown colouration having white bands. They are excellent pollinators, which give them an important ecological role in the places they inhabit. These bees are well known for their distinctive defensive behaviours and camouflage in dense forests. A more touching example is the specific behavioural response they exhibit against their chief predator *Oecophylla smaragdina* (weaver ant); when these ants are in close proximity, the bees produce adhesive barriers to obstruct the ant's path.

Defending a large social honey bee colony containing several thousands of workers requires simultaneous action of many individuals, which ideally involved communication between workers, enabling coordinated action and a fast response. *Apis florea* employ a defensive strategy called «piping», in which one individual emits an initial warning signal («piping»), which is followed 0.3-0.7 seconds later by a general response from a large number of bees («hissing»). With a fundamental frequency of 384 ± 31 Hz that lasts 0.82 ± 0.35 seconds, piping is audible to the human ear. Hissing is a noisy, broad band signal that is also audible to the human ear and produced by slight movements of the bees' wings. Individuals close to the piping bee are the first to hiss, which spreads rapidly to neighbors until an impressive coordinated crescendo is produced by the entire colony. Piping and hissing are accompanied by workers ceasing activity such as foraging dancing and departures from the colony.

A.florea demonstrates aggressive behaviors when competing for territory. Additionally, they compensate for their small size and limited flight range with more aggressive behavior since their flight range is also limited. This suggests that aggressive behavior is naturally selected for in *A. florea* because it facilitates defending flowers in a smaller territory.

Stingless Bee

Stingless or dammar bees are of smallest size compared to other honey-yielding bees (less than 5 mm). They belong to the family Apidae and subfamily Meliponinae. It consists of two genera Melipona and Trigona. Meliponinae includes eight genera, having 15 subgenera and more than 500 species. These bees are widely known as dammar bees in India (dammar is a resin from among dipterocarp trees) with additional local names commonly applied, e.g. "putka" in Sikkim and Nepal; "ngap siwor", "ngap hamang" and "ngap khyndew" in Khasi language; and "cherutheneecha" and "arakki" in Kerala. As the name implies, these bees can't sting as their stingers are highly reduced, but they try to defend their colony from intruders using their mandibles. The stingless bees are important pollinators of various food crops and can be domesticated. But the honey yield per hive per year is very low approximately 100 g. As in other regions where stingless bees occur, colonies can be kept in tree logs, wooden boxes and clay pots for harvesting small quantities of highly prized medicinal honey, wax and propolis, used for its household and curative properties. The materials used for nest building are mainly pure wax or cerumen, a mixture of wax and propolis, resins, plant fibres and clay.

Being tropical, stingless bees are active all year round, although they are less active in cooler weather, with some species presenting diapauses. Unlike other eusocial bees, they do not sting, but will defend by biting if their nest is disturbed. In addition, a few (in the genus *Oxytrigona*) have mandibular secretions, including formic acid, that cause painful blisters. Despite their lack of a sting, stingless bees, being eusocial, may have very large colonies made formidable by the number of defenders.

Stingless bees usually nest in hollow trunks, tree branches, underground cavities, termite nests or rock crevices, but they have also been encountered in wall cavities, old rubbish bins, water meters, and storage drums. Some beekeepers put them in bamboos, flowerpots, coconut shells, and other recycling containers such as a water jug, a broken guitar, and other safe and closed containers.

The bees store pollen and honey in large, egg-shaped pots made of beeswax (typically) mixed with various types of plant resin; this combination is sometimes referred to as "cerumen" (which is, incidentally, the medical term for earwax). These pots are often arranged around a central set of horizontal brood combs, wherein the larvae are housed. When the young worker bees emerge from their cells, they tend to initially remain inside the hive, performing different jobs. As workers age, they become guards or foragers. Unlike the larvae of honey bees and many social wasps, meliponine larvae are not actively fed by adults (progressive provisioning). Pollen and nectar are placed in a cell, within which an egg is laid, and the cell is sealed until the adult bee emerges after pupation (mass provisioning). At any one time, hives can contain 300–80,000 workers, depending on species.the nest cavity, includes the entrance tubes, is generally lined with of a mixture of secreted wax, plant resins ("propolis"), and other substances such as animal feces.

Himalayan Giant Rock Bee, *Apis dorsata lamboriosa*

Himalayan Giant Rock Bee, *Apis dorsata lamboriosa hive*

STINGLESS BEE, *Trigonula irridipennis*

STINGLESS BEE, *Trigonula irridipennis hive*

Table 1: Relative value of Different Honey Bee species

Particulars	Indian Bee	Italian Bee	Rock Bee	Little Bee	Stingless Bee
Habitat	Plains and Hills	Plains	Plains and Hills	Plains	Plains
Nature of Comb	Parallel	Parallel	Single Big Comb	Single Small Comb	Sac like Comb
Temperament	Gentle and easy to handle	Very gentle	Ferocious	----	----
Honey gathering	Good gatherer	Very good gatherer	Good gatherer	Poor gatherer	Very poor gatherer
Honey yield/ year	1-2.5kg[plains] 4-5kg[hills]	50-200kg	40kgs	0.5-1kg	60-180ml
Swarming	Prone to heavy swarming	Swarmless	Prone to heavy swarming	Prone to heavy swarming	----
Resistance to enemies	Helpless against wax moth	Guards hives except from wasps	Guards the hive	----	----
Stinging	Stings	Stings	Stings	Does not sting	Does not sting but bites
Domestication	Domesticated	Domesticated	Not domesticated	Not worth for domestication	----

CHAPTER 3

Honeybee Morphology and Physiology

Honey Bee External Morphology

Honey bees belong to the animal kingdom, and within it, to the group of arthropods (articulated legs), to the order of hymenopterans (membranous wings) and to the apidae family. Honey bees body is divided into three parts that are joined and that move each other: head, thorax and abdomen. Their external skeleton (exoskeleton) is composed of chitin, which gives them the stability that they need. It also protects the three large parts into which their body is divided, having a hard structure around head and thorax, and being extensible on the abdomen area.

Head: Honey bee head is a chitin box which has the shape of an inverted triangle. It houses their sight organs (simple eyes and compound eyes), two antennae and their mouthparts. It is attached to the thorax by a narrow and membranous neck.

Eyes: Three simple eyes, also known as ocelli, are located on the top of their head, between two compound eyes. They are covered with tactile hairs and have a very simple structure. Bees use ocelli to see at close range, and in almost dark conditions inside the hive. It has been verified that they are sensitive to the intensity of light and are used as photometers. The two compound eyes are formed by numerous hexagonal facets and each of them by thousands of simple eyes (queen has 3,000, workers 6,000 and drones 13,000). The shape of these facets is related to the kind of construction of combs. Their vision of colours varies with respect to human vision. They have more visual acuity on the ultraviolet side of the spectrum. On the red side they are practically blind. They see blue, yellow, green-blue and ultraviolet very well. They see red colour as if it were black and they confuse orange and yellowish green as if they were yellow. Although their visual acuity is lower than that of man, bee's eyes perceive 10 times more images in an equal time. They perceive polarised light, that is, the kind of light in which rays vibrate in a single plane.

Antennae: Two antennas emerge from the centre of the face, very close to each other. They articulate with the head by means of a membrane. Antennas are formed by a rigid part (scape) and a flexible part (flagellum) that is divided into segments. The portion that comes after the scape is called pedicel, which is a segment that is also part of the flagellum. The number of segments is 12 in queens and workers and 13 in drones. Antennas have numerous sensory organs like hair and pores. They are 3,000 in queens, from 3,000 to 3,600 in workers and 30,000 in drones. These organs are responsible for touch, hearing and smell. Bees' hair or hair organs are organs of touch and cover most of the antenna. Pores are funnel-shaped and are used for smell. If we make a cross section of the antenna and observe it under the microscope, we will clearly see inside it a network of nerves that serve as receivers and transmitters of sensations.

Mouthparts: The mouthparts of the honey bee belong to the chewing–lapping type, comprising the glossa, galeae and the vestigial labial palp, and they aid the bee in chewing pollen and lapping nectar. Labrum and mandibles are as in biting and chewing type of mouth parts. But mandibles are blunt and not toothed. They are useful to crush and shape wax for comb building; ingest pollen grains and other manipulative functions. Maxillolabial structures are modified to form the lapping tongue. The tongue unit consists of two galea of maxillae, two labial palpi and elongated flexible hairy glossa of labium. The glossa terminates into a small circular spoon shaped lobe called spoon or bouton or flabellum which is useful to lick the nectar.

When the bee is drinking nectar, the galeae and the labial palp form a sucking tube, within which the glossa produces a dipping motion with forward and backward movements. Current research shows that olfactory and gustatory perceptions begin at the mouthparts of a honey bee, where specific receptor neurons are located within specialized hairs, the sensilla . This implies that the swinging action of the joint-like articulations of the mouthparts when the mouthparts are engaged in physical activity may potentially help honey bees to capture more odor molecules for environmental perception.

Mandibles	They are blunt dumble shaped and are not toothed. They are not used for feeding but are useful for moulding wax into cells for comb (next) building.
Libium	The glossae are greatly elongated to form a hairy, flexible tongue. The glossa terminates into a small circular spoon shaped lobe called flabellum, which is useful to lick the nectar. Labial palms are elongate and four segmented.
Maxillolabial Structures	Maxillobial Structures are modified to form the lapping tongue. The tongue unit consists of the two galeae of maxillae, two labial Palpi and elongated flexible hairy glossa of labium
Feeding Mechanism	The galeae fit tightly lengthwise, against the elongated labial palps and they in turn roof over the elongated glossae (tongue) to form a temporary food channel through which saliva is discharged. The tongue (glossae) is trusted into flower, which gets smeared with nectar. It is then retracted between labial palps & galeae. Nectar is then squeezed by galeae and is deposited in the cavity formed by the paraglossae. Accumulated nectar is then drawn into oesophagus by the pharyngeal pump.

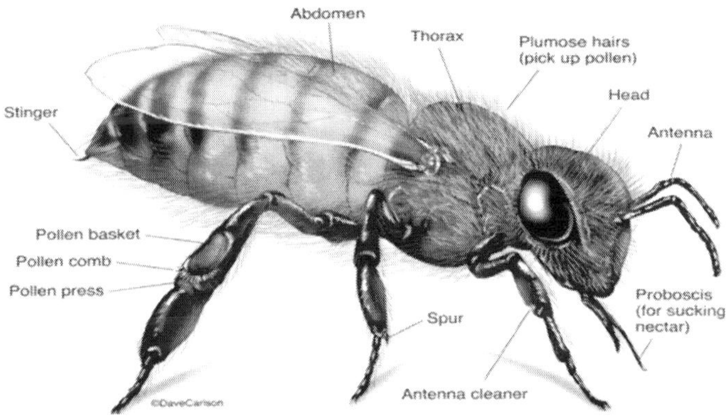

Honey Bee External Morphology

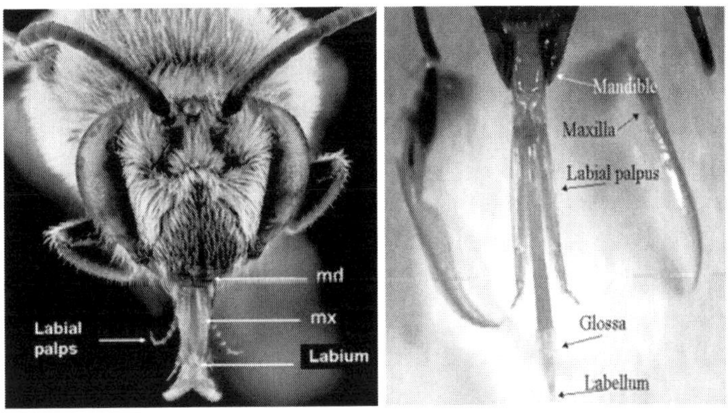

Honey Bee Mouth Parts – Chewing And Lapping Type

Thorax: The thorax is the center for locomotion and has three segments, each with a pair of spiracles for letting in air. Bees have 2 pairs of wings and three pairs of legs. The legs are very versatile, with claws on the last tarsomere, allowing bees to have good grip on rough surfaces (tree trunks etc), but also with a soft pad (arolium) to allow bees to walk on smooth surfaces (leaves or even glass!). There are also special structures on legs to help bee get more pollen.

Legs: Each pair of legs differs in size and shape from the other two pairs and is jointed into six segments, with a pair of claws at the tip which helps the insect to cling to surfaces. The leg can be flexed at any of the six joints. Its primary function is to help the bee to walk and run, but various parts also serve special purposes other than locomotion. For example, the brushes on the inner surface of the fifth segment, (the tarsus) of the two front legs are used for sweeping pollen and other particles from the head, eyes and mouth parts. The same tarsi of the mid-legs serve as brushes for cleaning the thorax, while the spines found at the end of the fourth sections (tibiae) are used for removing the pellets of pollen and for cleaning the wings. Important parts to note on the legs are the antenna cleaners on the front legs, wax picking on the middle legs and the pollen baskets on the hind legs.

1. **Forelegs:** Bees, like other hymenoptera, have a special structure on their forelegs (strigilis) to clean their antennae. The antenna cleaner of the honeybee is composed of a deep notch on the inner (ventral) surface of the proximal end of the basitarsus and a large spur, which is movable, inserted at the inner angle of the distal end of the tibia. The notch bears on its posterior end a semicircular comb of fine, closely set hairs. The spur consists of a trunk with an apex and two lobes (vela), the larger of which is directed posteriorly, the smaller (spatulate accessory lobe,) anteriorly.

   Honeybees (*Apis mellifica*) invariably clean their antennae with the ipsilateral foreleg. They hardly ever clean both antennae at the same time even if both are soiled. In this case the bee cleans first one, then the other antenna. Almost invariably honeybees clean their antenna with only one stroke, then turn to other behavior. Immediately before the cleaning stroke the antenna is lowered, and the first leg is lifted above the antenna. Often, though not always, the head is turned toward the cleaning leg, but it is not lowered. The antenna is put into the strigilis, approximately in the region of the pedicel or the first flagellar segment. By bending the basitarsus against the tibia the antenna cleaner is closed around the flagellum. The subsequent scraping movement is directed back- and downwards, i.e. distal with regard to the antenna. The strigilis thus only cleans the flagellum, not its base. While performing the cleaning stroke, the bee presses the scape of the antenna against its head (gena). Sometimes the stroke is reinforced by a counter rotation of the head. The distal part of the flagellum, which has not yet passed the strigilis is bent upwards. This is especially evident toward the end of the stroke.

2. **Middle legs:** Honey bee workers have four pairs of special wax-secreting glands on the undersides of their abdomens. From these glands, they secrete liquified wax, which hardens into thin scales when exposed to the air. As the worker bee ages, these glands atrophy and the task of making wax is left to younger bees. After the soft wax hardens, the worker bee uses stiff hairs on her hind legs to scrape the wax from her abdomen. She passes the wax forward to her middle legs, and then to her mandibles. Hence, the middle legs are referred as wax picking legs.

3. **Hind legs:** The pollen basket or corbicula (plural corbiculae) is part of the tibia on the hind legs of certain species of bees. They use the structure in harvesting pollen and carrying it to the nest or hive. The corbicula is a polished cavity surrounded by a fringe of hairs, into which the bee collects the pollen; most other bees possess a structure called the scopa, which is similar in function, but is a dense mass of branched hairs into which pollen is pressed, with pollen grains held in place in the narrow spaces

between the hairs. A honey bee moistens the forelegs with its protruding tongue and brushes the pollen that has collected on its head, body and forward appendages to the hind legs. The pollen is transferred to the pollen comb on the hind legs and then combed, pressed, compacted, and transferred to the corbicula on the outside surface of the tibia of the hind legs. In *Apis* species, a single hair functions as a pin that secures the middle of the pollen load. Either honey or nectar is used to moisten the dry pollen, producing the product known as bee pollen or bee bread. The mixing of the pollen with nectar or honey changes the color of the pollen. The color of the pollen can help identify the pollen source.

Wings

Honey Bees possess a pair of forewings that are similar in length to the abdomen and a smaller pair of hindwings. The 2 vs 4 ignorance is somewhat understandable when considering that when bees are not in flight the larger forewings cover the hind wings in the relaxed position along the abdomen, and the wings function as a single pair during flight activity. This coupled movement is facilitated by a series of hooks, known as *hamuli*, on the leading edge of the hind wings that fit into grooves on trailing edge of the forewings allowing the forewings and hind wings on each side of the body to function as a single surface. The wings are powered by the longitudinal and vertical flight muscles in the thorax. Alternating contractions of the vertical and longitudinal muscles raise and lower the wings.

Bees are capable of flying at speeds up to 15 mph and carrying nectar loads that approach their own body weight. In addition to these feats of strength, they are also capable of delicate maneuvering and hovering while they approach flowers. These amazing aerial abilities are of course made possible by their wings, which at a glance seem undersized for the task. In addition to propelling themselves through the air, the wings are also used to move air while remaining stationary. The ability to move air throughout the colony is critical to maintaining appropriate internal temperature and humidity as well as evaporating moisture from nectar and distributing pheromones within the colony.

The wings themselves are composed of three layers: a transparent membrane on top and bottom supported by a network of veins that carry hemolymph (bee "blood"), nerves, and breathing tubes throughout the wings as well as provide structural support. Honey bee wings exhibit a relatively simple pattern of venation compared to other more primitive insects. Venation patterns can be used to distinguish between subspecies (or races) of honey bees. This requires precise measurements of different sections of the wing and is being increasingly done through automated image processing algorithms.

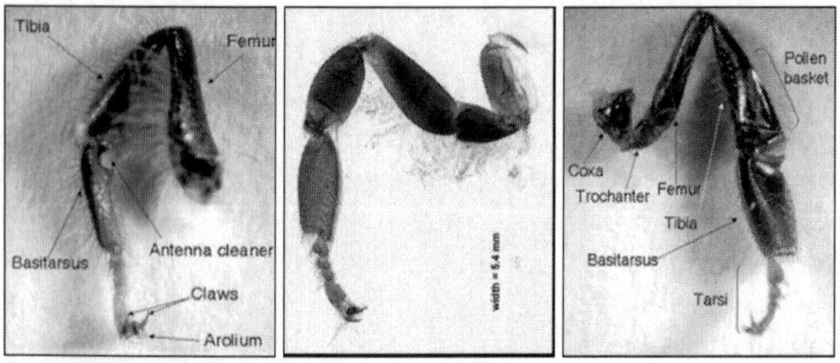

Front Leg-Antenna Cleaning   Middle Leg – Wax Picking   Hind Leg –pollen Basket

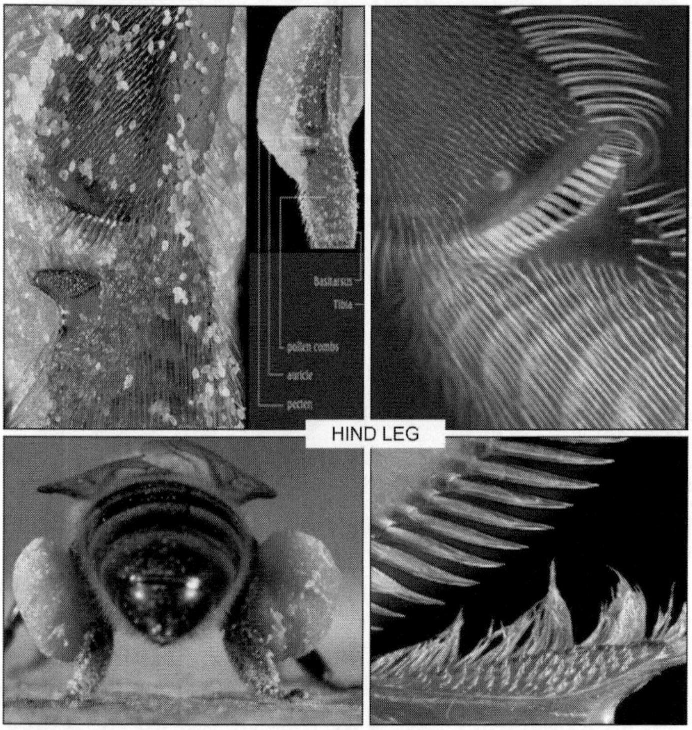

HIND LEG

Despite their durability, honey bee wings are only capable of a finite number of flight miles, with estimates of this upper bound being approximately 500 miles. Close observation of older bees in the colony will often reveal wings that are tattered around edges. As adult bees progress through their lifespan, they take on different tasks and roles in a process known as temporal polyethism with foraging typically being the last activity. When you see an older bee, which can be identified by diminished body hair in addition to tattered wing margins, you're observing a bee that is near the end of its life. Individual bees only occupy the role of forager for 2-3 weeks, covering hundreds of collective miles in that time. When you see an old bee with tattered wings it is documentation that she has exhausted her physical capabilities in service to the colony and will soon be replaced by a younger bee that transitions from a house bee to a forager.

Average worker honey bee beats its wings at the rate of 208 to 277 beats per second. That adds up to 12,480 to 16,820 beats per minute. Really, though, there's so much variety in the insect world that, in a wing-flapping contest a fly would beat a bee. Some flies beat their wings up to 1000 times per second. Butterflies couldn't qualify for the contest. They beat their wings only 5 to 20 times per second.

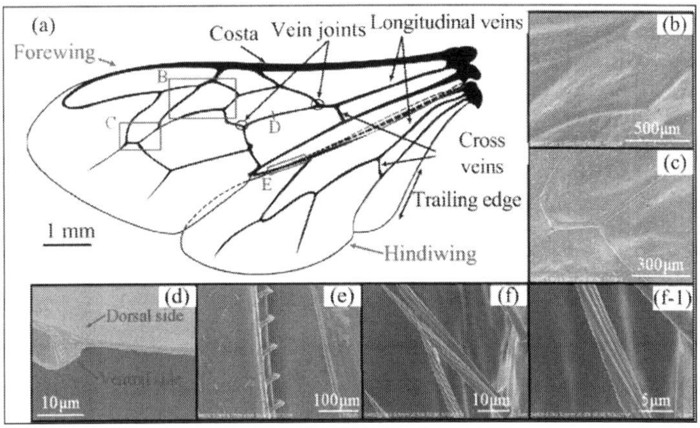

Honey Bee Wing Venation

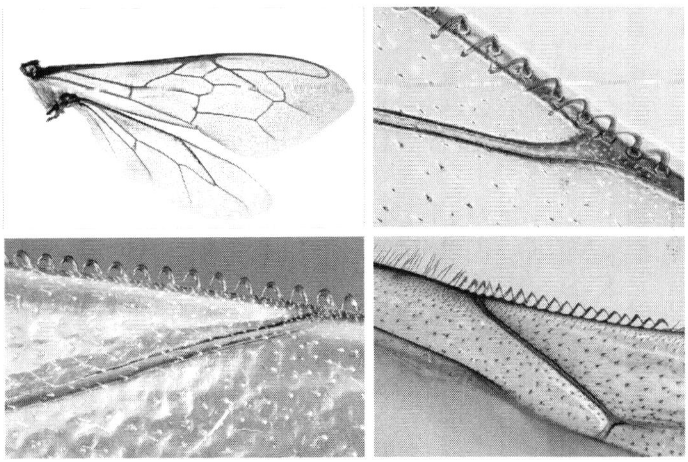

Honey Bee Wing Coupling - Hamulii

Aeroplanes can fly because of a careful balance of four physical forces: lift, drag, weight, and thrust. The lift force must balance its weight, and thrust must exceed its drag, to make flying possible. Planes use wings for lift and engines for thrust. Drag is reduced thanks to a streamlined shape, and lightweight materials. The wingspan of a plane is large enough to satisfy the lift equations for flight, so they don't need to flap. But the small wings of a bee compared to its relatively fat body are not. A regular Boeing 747 plane can also take off at roughly 184 mph, whereas bees do not reach anywhere near that speed. Due to low speeds, and the high amount of drag when bees flap their wings, it might look like they shouldn't be able to fly. In reality, they simply fly in a completely different way. One study from 2005 helped explain the way bees get themselves off the ground. The scientists compared bees to fruit flies, and found that a fruit fly has one eightieth the body size and flaps its wings at a rate of 200 times per second. In comparison, honeybees flap 230 times per second. This was surprising because smaller insects generally have to flap their wings faster to compensate for decreased aerodynamic performance. To further complicate things, bees are also often carrying pollen and nectar, which sometimes weighs as much as their entire bodies.

Abdomen: The abdomen of the worker appears to consist of 6 segments. However, it must be remembered that the first true abdominal segment is closely associated with the thorax. The segments of the abdomen readily are visible in the worker or queen which number II to VII. Segments VIII-X are concealed within segments VII. They are highly modified and are difficult to recognize as entities. The drone appears to have an additional segment which is composed often 8th tergite and the 9th sternite. In both sexes, the xth segment is reduced to a small lobe which bears the anus. The abdomen bears 7 pairs of spiracles apart of those on the propodeum. It also bears 4 pairs of wax glands, a dorsal scent gland, and in the worker and queen, the sting and its associated structures. It contains most of the visceral organs.

The Wax Glands: The wax glands are located on the ventral surface of abdominal sternites IV to VII. The anterior portion of each of these sternites bears two transparent, polished surfaces known as mirrors. The wax glands are located inside the abdomen over the mirrors. The wax is secreted as a liquid and forms a thin flake in each wax pocket between the mirrors and the posterior, under lapping portion of the preceeding stemite.

The Dorsal Scent Gland: Located under the base ofthe VIIth abdominal tergite is the dorsal scent gland, or gland of Nassenoff. Mclndoo has described the structure as being composed of unicellular epidermal gland cells. It is believed that the secretion of the scent gland passes through the epidermis by tiny ducts opening to the exterior. According to Mclndoo, the scent gland of the queen is larger than that of the worker. This has been disputed by Jacobs, who stated that the scent gland is lacking in the queen and in the drone. The scent is used as a means of communication between the individual bees.

The Sting: The sting of the worker and queen bees is a modified ovipositor. Instead of being used to prepare a suitable site for eggs, as it does in many other insects, the sting is used to inject poison. The sting and most of its associated organs are contained in a special cavity or sting chamber. The sting consists of a dorsal style or sheath and two ventral lancets. The stylet is enlarged basally. It has a pair of projecting condyles and each lancet possesses a dorsal groove. The lancets can slide freely along the condyles but are held closely together. The stylet and lancets form a cavity, the poison cannel, through which poison is injected when the bee stings. The lancets possess strong recurved teeth in the worker. While the sting of the queen possesses barbs, they are not so well developed and the sting is more strongly attached. Hence, the queen does not lose her sting, where as the sting and associated organs of the worker are tom from the body in the process of stinging. Associated with the sting are three pairs of plate, the quadrate which is the largest, the triangular which is the smallest, and the along which is intermediate in size, a pair of fresh lobes formerly known as the sting palpi, but now referred to as the sting sheath, and the poison sac, poison gland, and alkaline gland.

Wax Secretion by Honey Bees

Honey Bees Constructing Bee Hive Using Wax

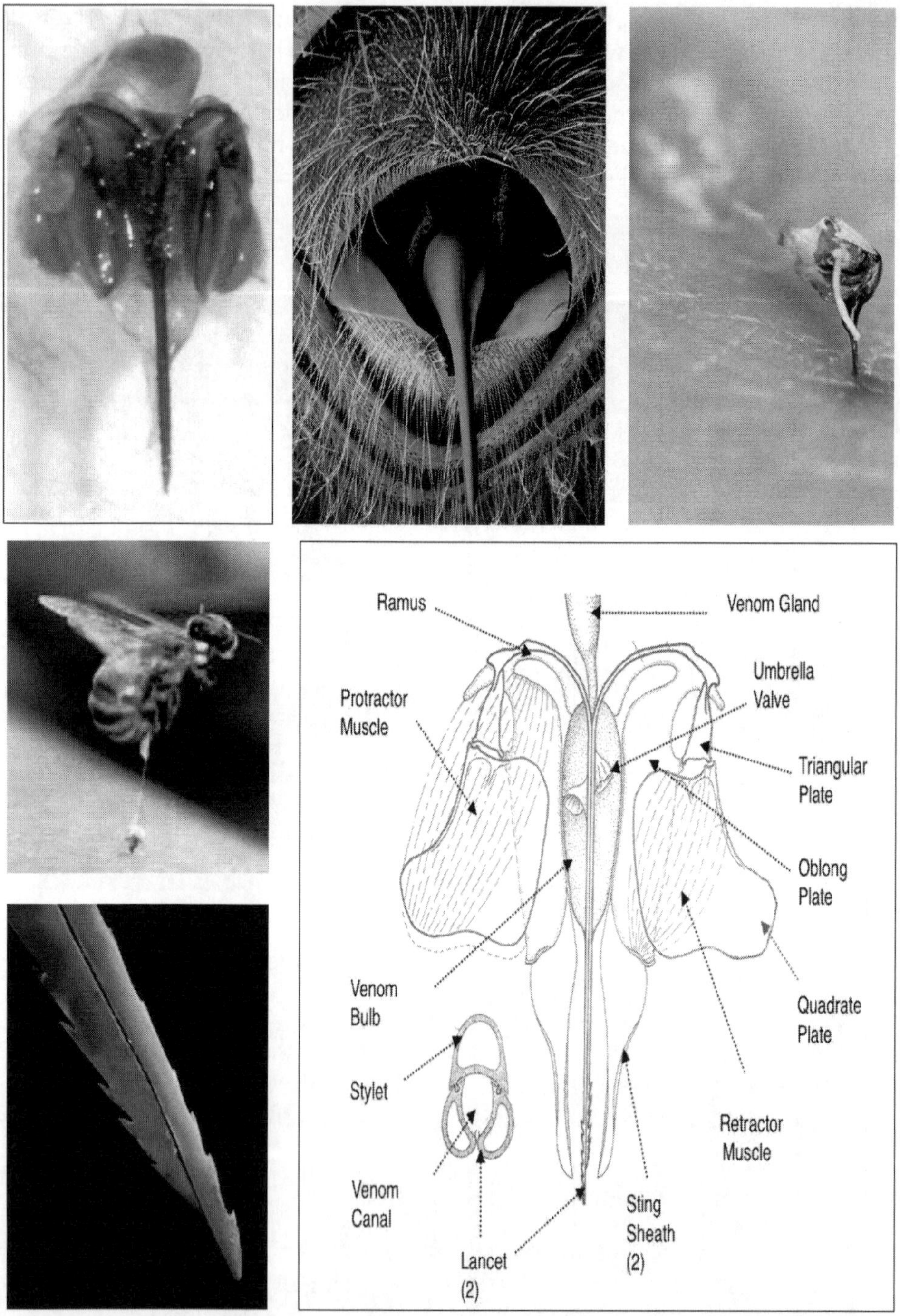

Honey Bee Sting

All these structures function in the process of stinging. It should be noted that the sting is capable of independent motion once it has been inserted in an enemy. By muscular action, the sting and poison are driven deeply into the wound. In the process of stinging, the bee curves the abdomen ventrally, extrudes the sting, and thrusts it into the victim. The barbs on the lancet hold the sting in the wound. By muscular action, first one and then the other lancet are 50 driven deeper into the wounded. Unless the sting is removed, it is soon embedded for its entire length. The poison, in the meantime flows. To remove a sting, the fingernail or a knife blade should be used to scrape this organ from the flesh. If the sting is grasped by the fingers or by tweezers, usually more poison is squeezed into the wound. Beekeepers generally scrape out the sting by rubbing their hands or arms against their clothing or by scathing the area stung.

Honey Bee Anatomy - The internal morphology deals with different organs or structural parts that assemble to facilitate proper metabolism in the honey bee system. There are various systems are: Digestive system, excretory system, nervous system, circulatory system, respiratory system, reproductive system and special organs.

Digestive System

The digestive system is composed of three main sections, the **foregut, midgut**, and **hindgut**. Interestingly enough, the three sections of the digestive tract form separately during bee development. The foregut (first third of the digestive tract) and hindgut (last third of the digestive tract) form as invaginations from either end of the developing bee. The foregut is composed of the **mouth, oesophagous**, and **crop** of the honey bee. Food enters the digestive tract through the mouth and travels down the esophagus and into the crop. The esophagus is simply a tube that runs from the mouth in the head, through the thorax, and into the crop in the abdomen. The crop, or **honey stomach** as it is called sometimes by beekeepers, is a spherically shaped organ in the abdomen that serves as a site for food storage, as a storage place for nectar bees collect from flowers and fly back to the hive, or as an initial site for the digestion of food in the bee. The crop can expand significantly when it is full of honey or nectar, so-much-so that the abdomen swells. The foregut and midgut are separated by a valve called the **proventriculus** which is located at the end of the crop. This valve can grind and pulverize food particles (such as pollen) and filter pollen out of the crop contents. Food passes through the proventricular valve and into the bee's midgut or **ventriculus** .

The midgut is the primary site of enzymatic digestion of food and absorption of nutrients. It is not lined by cuticle but rather is lined by the peritrophic membrane. This membrane likely protects the digestive cells (the cells that line the internal surface of the midgut) while allowing absorption of the nutrients straight into the hemolymph. Because the midgut is somewhat permeable, being so because of its function as the site of nutrient absorption, this is where many viruses and other bee pathogens enter the hemolymph. This is true especially for the Nosema pathogens (*N. apis* and *N. ceranae*) and some viruses.

Next along the digestive tract are the Malpighian tubules. These occur at the end of the midgut and are, essentially, spaghetti-like extensions of the tract that float freely in the bee's body cavity. The Malpighian tubules extract waste products from the hemolymph. They produce uric acid granules and help with osmoregulation (water management) within the bee.

The hindgut, or final section of the digestive system, is composed of the ileum and rectum. The ileum, sometimes called the small intestine, is a short tube that connects the midgut to the rectum. The rectum is important for the absorption of water, salt, and other beneficial substances prior to waste excretion. There are small areas on the rectum called rectal pads. These sections reabsorb >90% of the water that was used by the Malphighian tubules to collect waste. The latter is an important function. Bees, like most insects, try to retain as much moisture as possible from the food they eat. Thus, they do not excrete nitrogenous wastes in a urine-equivalent as humans do. Instead, they reabsorb much of their water and tend to defecate moderately liquid to dry feces. Uric acid solids and other unused foodstuffs, such as the shells of pollen grains, are excreted as relatively solid feces.

The Excretory System: The waste products of metabolism of the bee are removed principally by the Malpigian tubes. However, both $C0_2$ and some water pass off through the tracheae. Some investigators also have claimed that nitrogenous wastes are accumulated in the urate cells which are associated with the fat body. Uric acid deposits, together with certain pigments, are also found in the cuticle of some insects. The pericardial nephrocytes are also believed to act in excretion. The Malphigian tubes are small, tubular structures which extend throughout the abdominal cavity. They number about 100 in the adult. The wall of each is one cell in thickness. The tubes remove wastes from the blood and these are discharged into the alimentary canal at the junction of the midgut and the hindgut.

The Nervous System: Insects in general have three types of nervous systems: (i) the central nervous system, (ii) the stomatogastric or sympathetic nervous system, and (iii) the peripheral nervous system. We have more detailed information on the central nervous system of the honeybee than we have on the other systems. The central nervous system consists of a dorsal brain and a row of ventral ganglia joined by paired longitudinal connectives. There are two large nerve masses in the head, the brain and the sub-oesophageal ganglion. The oesophagus passes between these two masses, which are joined by circumoesophageal connectives. Both nerve masses of the head are compound ganglia. The brain is composed ofthree more or less clearly defined parts. These include the protocerebrum, the deutocerebrum, and the tritocerebrum. These nerve masses innervate the antennae, the compound and simple eyes and the labrum. By the frontal commissures, the brain is connected to the frontal ganglion and the stomatogastric nerve which extends posteriorly along the dorsal wall of the pharynx. The drone possesses the largest brain, but this is due primarily to the great development of the optic lobes. As far as the true brain is concerned, the worker possesses the largest. The sub-oesophageal ganglion innervates the mandibles, the maxillae and labium.

Posteriorly, the suboesophageal ganglion is united to the first thoracic ganglion by long paired connectives. The stomatogastric nervous system is concerned with the innervations of the digestive, circulatory and reproductive systems. The peripheral nervous system is concerned primarily with the innervations of sensory cells in the integument. The ventral nerve trunk consists of seven nerve masses or ganglia. The first is located in the prothorax and its principal function is the innervations of the first pair of legs. The second ganglion of the ventral nerve trunk lies approximately at the juncture of the meso-and meta-thoraces. This is a compound ganglion and innervates the meso- and meta-thoraces, the propodeum and the second true abdominal segment. In the abdomen, there are five ganglia which innervate abdominal segments III-VII. Since, the posterior ganglion innervates all the segments posterior to it must be a compound ganglion composed of ganglia representing segments 7-10. As in other insects, the coordination of the muscles does not depend on the brain. Each ganglionic mass appear to have more or less independent control over the portion of the body it innervates.

The Circulatory System: The circulatory system consists of a single pulsatile organ, the heart aorta. In addition, the flow of blood is also regulated by dorsal and ventral diapharagms. The 5-chambered heart is found in the dorsal part of the abdomen. Each chamber of the heart possesses a pair of ostia or slits which permit blood to enter the chambers. The lips of the ostia project anteriorly into the chamber ahead and act as valves to prevent the blood from flowing backward. The aorta is a tubular continuation of the heart. It has no chambers of ostia. The aorta extends through the thorax into the head where it divides and Fig.3.8 Circulatory system of honey bee opens below the brain. The blood is pumped anteriorly from one chamber to another by the rhythmic contraction of the heart. It passes through the aorta and pours out into the head near the brain. It then flows back through the thorax to the abdomen. The pulsation of the ventral diapharagm force, the blood backward and probably upward. The dorsal diaphragm forces the blood anteriorly. The blood of the bee is a pale yellowish or somewhat amber-colored liquid. It consists of a fluid in which various form of blood cells or haemocytes occur. These cells are more closely comparable with the white blood cells or leucocytes of the vertebrates than with the red blood cells. The principal functions of the blood are the transportation of digested food products and the removal of waste from the cells in all parts of the body. These wastes are then eliminated principally by the Malphigian tubes. In the bee, oxygen is not carried by the blood cells, as it is in the vertebrates.

The Respiratory System: The respiratory system consists of long, tubular, branched investigation of the ectoderm known as tracheae. The arborescent ending of the tracheae, the tracheoles, come into contact with practically all the body cells. Hence, the exchange of oxygen and CO_2 occurs without any necessity for blood as a carrying medium. Air is taken into the body through the spiracles. It soon enters the air sacs, which are thin-walled expansions of the longitudinal tracheae. From the air sacs, transverse tracheae extend dorsally and ventrally.

The trachea is subdivided to form tracheoles which are closely attached to the various organs and which end blindly. In some insects, the ends of the tracheoles are filled with liquid, which is believed to facilitate the exchange of oxygen and co2. Bailey has indicated that gaseous exchange is modified by the activity of the bee. He found that inactive bees inhale and exhale Fig.3.9 Respiratory system of worker bee. Aorta Brain Heart / Dorsal diaphragm /Ventral diaphragm Antennal vesicle through the first thoracic spiracles, but that active bees producing a high $C0_2$ tension inhale through the prothoracic and abdominal spiracles and exhale through the propodeal ones.

The Reproductive System: The reproductive organs produce sperms or ova from which succeeding generations develop. In many animals, these are comprised of external and internal organs, but in the bee they are practically all internal. Since the structures of the two sexes are markedly different, the discussion of these systems must be considered independently.

The female reproductive organs: The reproductive organs of females are comprised up of multiple structures. A laying queen's ovaries are pear shaped in general. The eggs develop within the 160–180 tubular ovarioles which make up each ovary. An oviduct is situated at each ovary's posterior end. The median common oviduct, which unites the two lateral oviducts, forms one and connects into the vagina. This then leads to the bursa copulatrix, which is located close to the sting's base and opens to the outside. The spermatheca is a spherical organ where the drone's spermatozoa are held until insemination. It is reached by a tubular spermathecal duct that originates from the vagina's dorsal wall.

The terminal segment of the spermathecal duct gives rise to paired spermathecal glands. Primary reproductive cells in the ovariole tips are the origin of the ova. The ovariole develops from the developing ova. Nutritive cells, commonly referred to as nurse cells, accompany each ovum and provide nourishment. The ovariole's follicle cells secrete a thin shell termed as the chorion to protect the fully formed eggs. With the exception of a small opening at the anterior end of the egg called the micropyle, the egg is completely enclosed in shell.

This widens an entrance for the spermatozoa to pass through and fertilize the egg. The queen bee has 73 polytrophic ovarioles in each of her ovaries. The recently emerged queen bee's ovary is tiny and lacks stages of the oocyte that are developed. Nurse cells appear to be exhausted in the aged ovary, exhibiting black granules. The worker bee's ovary contains one to twelve ovarioles. The worker bee's reproductive organs have a calyx that is a solid structure with an irregularly shaped intima lining the aperture. There is a layer of circular muscle around its epithelium. The synchronized ovulation of the ovaries has been linked to an inconsistent egg-laying behavior in worker bees. The spermathecal gland's glandular epithelium has two distinct types of nuclei: endothelium and big. Each cell has a duct which culminates in the gland's lumen. It has been examined how the two domesticated species of *Apis* have evolved and their affinities.

The male reproductive organs: There are numerous intricate structures that make up the male reproductive organs. The testes resemble little kidneys and are flattened. The testicular follicles, or tubules, are many in the testes and are where the male sex cell develops. A big, tubular, seminal vesicle is the result of the follicles opening into the vas deferens. After being held in the vesicle for a short while, the mature spermatozoa finally descend into the penis bulb. Two sizable mucosal glands are connected to the seminal vesicles, and the vesicles empty into these glands. One tubular ejaculatory duct runs from the mucous glands to the penis bulb.

A ventral ridged cervix, a terminal bursa, and a dorsal fimbriated lobe are some of the features that make up the penis. There are two dorsal horns on the bursa's dorsal wall. The male reproductive system's external aperture is located above the ninth sternite and beneath the anus. The penis is everted into the female's vagina during mating, and the spermatozoa travel to the spermatheca via the spermathecal duct. Following the union of the sexes, the male penis is removed from the body, while the penis bulb momentarily stays in the female genital tract.

The drone mates and then dies. One of the wonders of bee physiology is the way the spermatozoa are fed and maintained for several years in the spermatheca of the queen. The spermatheca has an abundant supply of trachea, which removes gasses and water vapor from the cells and supplies them with oxygen. In addition, it has a pump and two spermathecal glands. Together with the spermatheca's lumen and the duct that joins it to the common oviduct are these glands. It appears that the glands and pumps keep the spermatheca turgid and regulate the spermatozoa's entrance and release.

The spermatheca will not collapse if the duct flowing from it is severed between the pump and the oviduct; nevertheless, if the duct is destroyed between the pump and the point where the two glands unite, the spermatheca will remain rigid. Four to seven million spermatozoa can be found inside the supermatheca. They are discharged in modest quantities to fertilize the eggs that will give rise to worker larvae. It is unclear how the queen regulates their release. Seldom does a typical mated queen make the error of putting fertilized eggs inside a worker cell. It is rather easy to demonstrate that the type of egg laid is independent of the size of the cell.

When a freshly mated queen is given only combs made of drone cells, she will often deposit an egg in each cell that will eventually develop into a worker bee. There is a valve fold in the vaginal floor that, when lifted, would cause the egg to pass right in front of the aperture where the spermathecal duct opens into the queen's vagina. The smaller end of the egg is placed first and has a modest curvature. The micropyle, or opening, in the egg would come into touch with the duct opening in the vaginal roof due to this curvature and the valve fold, allowing the spermatozoa to do so. The egg would not come into touch with the duct opening if the valve fold is not lifted. It is very likely that some stimulus allowing the queen to lay a worker or drone egg whenever she pleases is directly associated with the operations of the spermathecal pump and the valve fold. Laidlaw and others have shown that in order to successfully artificially inseminate a queen, this valve fold must be adjusted.

Glands : The head contains the brood-food gland or hypopharyngeal glands. They are made up of two lengthy ducts that have 500 or more sac-like secreting bodies linked to them. It is commonly understood that these glands secrete the royal jelly or brood nourishment. On the other hand, other researchers suggested that the labial glands might also have a role in the production of these compounds. There are two groups of labial, or salivary, glands: the "post-cerebral glands" are located in the skull, while the second set is located in the thorax. It is widely accepted that the synthesis of enzymes including lipase, diastase, invertase, and proteinase is this organ's principal role. The mandibular glands are located in the head, directly above the mandibular bases. Their function is not entirely clear. The explanation that is most commonly accepted holds that they release a chemical that changes the characteristics of wax to make it ready for comb construction. They might also soften the pupal cocoon, according to another notion. The outermost layer of the rectum location for rectal pads. They stretch lengthwise and number six. The term "rectal glands" applied to them. Although their exact purpose is unknown, they might help keep the bee's water balance stable.

Oenocytes : The nucleus of the big, densely stained oenocyte cells is highly stained. Usually, they are connected to the fat cells. Their precise purpose is unknown. According to Snodgrass, there might be a connection between them and wax secretion. Some have proposed that the cells secrete enzymes essential to the bee's physiology.

Johnston's organ: This is positioned in the pedicel of antennal. Honeybees use Johnston's organ (JO), which is housed in the pedicel of each antenna, to sense vibrations in the air. Using intracellular recordings in conjunction with intracellular labeling, it is known that two types of vibration-sensitive interneurons with arborizations are present in the major sensory area of the JO, namely, the dorsal lobe-interneuron 1 (DL-Int-1) and dorsal lobe-interneuron 2 (DL-Int-2). The three-dimensional pictures of the individual neurons were registered into the conventional map of the honeybee brain in order to visualize overlapping areas between the branches of these identified interneurons and the JO sensory terminals.

Scolopophorous organs: These intricate structures are found in the honeybee's head and legs. They were once thought to have a chordotonal function, although more recent researchers have cast doubt on this theory. According to Snodgrass, they might register how tense muscles affect the body wall.

Endocrine organs: Hormones are chemicals that control the growth, moulting, and other developmental processes of bees and most other insects. These are produced by a variety of organs known as endocrine glands. Furthermore, a collection of neurosecretory cells found in the brain's paired intercerebralis release these chemicals. The corpora cardiaca, corpora allata, and maybe the thoracic glands are among the bee's endocrine glands. During the preinnago stage, the latter are discovered, and they vanish during pupation. They are located between the first and second ganglia of the ventral nerve cord at the sidewalls of the alimentary canal in the thoracic area.

The adult bee's corpora cardiaca are tiny organs situated above the area where the throat and oesophagus converge. Nerves that stretch to the corpora allata connect them to the brain. There is evidence to suggest that neurosecretory substances from the heart muscle may seep into the aorta. The globular corpora allata are situated next to the oesophagus's sidewalls. They have a close relationship with the heart muscle. The specific roles played by each of the honeybee's endocrine glands are not well understood. The corpora allata of other insects under study secretes a juvenile hormone that prevents the development of immature stages between mouths.

Nonetheless, it is known that the corpora allata of mature honey bees vary in size based on the activity level and the bees' age. This has given rise to the hypothesis that the corpora allata's secretion controls the honeybee's overall metabolism. It is thought that the brain's neurosecretory cells, as well as the corpora cardiaca, supply the hormone that controls growth and development. According to certain data, the brain's neuro-secretory cells may secrete a hormone that affects the thoracic glands, which in turn creates a hormone that controls growth and development.

The queen mandibular odorants, (2E)-9-oxodecenoic acid (9-ODA), (2E)-10-hydroxydecenoic acid (10-HDA), and (2E)-9-hydroxydecenoic acid (9-HDA), are picked up and memorized by the drones of Apis cerana and Apis florea. Apis cerana drones responded more strongly to 9-ODA than Apis florea drones did to 10-HDA. Both species' drones, though, reacted to the 9-HDA, with the Apis florea drones responding more frequently. The study's findings indicate that, in contrast to 10-HDA in Apis florea queens, 9-ODA is a significant sex pheromone component in *Apis cerana*.

CHAPTER 4

Honey Bee Biology and Colony

A. Honey Bee Biology

One of nature's most brilliant products is the honeybee. One of the best characteristics exhibited by honeybees is eusociality, which includes their caring for their young in a cooperative manner and possessing other sophisticated defense and communication systems. In addition to an adult stage, honeybees go through three developmental phases: egg, larval, and pupa. Three castes exist at the adult stage: thousands of workers, hundreds of drones, and a single queen. Because honey bees have a haplodiploid sex determination mechanism, the worker is an unfertilized female capable of generating only males, while the drone is a male. The queen is a fertile, functioning female that can create both males and females.

Their caste is established by the food they are fed during the larval stage; nurse bees feed queen larvae royal jelly at this time. It has recently been discovered that the larvae's development into queens is caused by the 57 kDa protein royalactin, which is found in royal jelly. Like with the majority of Hymenoptera, fertilized bee eggs become females, while unfertilized eggs become males. Sex is controlled by alleles at one or a few loci and heterozygosity at the sex determining locus (or loci) produces females.

Since heterozygosity is impossible, development without fertilization, or with a haploid number of chromosomes, results in males. Some diploid eggs produced by inbreeding are homozygous at the sex-determining loci, resulting in the production of diploid males. These males are typically not useful for reproduction because they have a low sperm cell count, a short lifespan (*Apis* larval mortality), and the propensity to produce triploid offspring that are incapable of reproducing. Therefore, hapiodiploid is the process that determines sex in honeybees.

A female retains sperm cells in her spermatheca after mating, typically receiving an endless supply. Then, when the egg travels through the oviduct, she can regulate the sex of each egg by releasing or not releasing sperm cells from the spermatheca. This design allows the female (of species where females are larger than males) to lay male-producing eggs in small cells and female-producing eggs in large cells with more resources. Male-producing cells in *Apis* are larger than worker-producing cells, and the size of the cell determines whether the queen fertilizes each egg or not. The males of *Apis* are larger than the workers.

Additionally, in burrows where the cells are built in series, the female bee can lay her eggs for the males in cells close to the entrance, allowing the offspring to go without disturbing the slower-developing females. A female bee's lifetime egg production ranges from eight or less for certain isolated species to over a million for queens of highly gregarious species. Solitary bee females tend to their few offsprings by choosing suitable nest sites, building nests, building and supplying brood cells, and identifying each individual offspring's appropriate sex. Thus, the limited reproductive capacity of many solitary bees is made possible by this concern to the well-being of progeny.

Almost all bee species have long, softly curved eggs that are white in color and covered in a soft, membrane-like covering called a chorion, or "shell," which is often placed on top of or, in rare cases, inside the food mass that the larvae will consume. Nonetheless, the eggs are laid with little to no associated food in bee species that feed their larvae gradually, such as Apis Bombus and the majority of Allodapini. Generally speaking, eggs are medium in size; however, in highly social bees, which lay numerous eggs in a single unit of time, and in Allodapula (Allodapini), which lays eggs in batches, multiple eggs at roughly the same time, eggs are much smaller.

Moreover, many cleptoparasitic bees have tiny eggs that are hidden in the walls of the brood cells of their hosts. These eggs are frequently very specialized in shape and may include an operculum, which allows the larva to emerge. Many bee species' mature larvae build cocoons, mainly during the larval defecation stage. The majority of short-tongued bees no longer spin cocoons, and they are frequently shielded by the mother bee's secretory cell line.

As holometabolus insects, honeybees go through a full metamorphosis. There are all four developmental stages: egg, larva, pupa, and adult. The tables show the length of the life cycles of the western bee, *Apis mellifera*, and the Indian bee, *Apis cerana*. The queen returns to her colony to begin laying eggs after her nuptial flight. A short, delicate white tubular egg with a slight curvature on one side is adhered to the bottom of a comb cell by the mother queen using glue or stick. A little white larva emerges from this egg and devours the nourishment that the nurse bees provide. It weights several times its initial weight when completely nourished. The worker bees cap the cell with a waxen covering. After finishing its meal, the fully grown larva spins a cocoon inside its cell and goes into a resting phase, when it changes into a pupa. The body's major structural alterations come at the expense of the fat that the larvae have accumulated there. Since the pupa has fully formed mouth parts, legs, wings, and other body appendages, it resembles an adult bee in appearance. After the cell's waxen cover is removed and the cocoon is cut, the adult bee emerges. Honey bees have a haplodiploid sex determination system. Fertilized eggs yield female offspring, while unfertilized eggs, or parthenogenicity, yield male offspring, or drone.

Table 1: Life cycle of different caste of *Apis cerana india*

Categories Caste	**Duration of stage (Days)**			
	Egg	**Larva**	**Pupa**	**Total**
Queen	3	5	7	15
Worker	3	5	10	18
Drone	3	7	14	24
Life cycle of *Apis cerana* india during different seasons (in days)				
Seasons	Egg	Larva	Pupa	Total
Spring (March-May)	3	6	10	19
Summer (June –August)	3	6	12	21
Autumn (September-November)	3	6	10	19
Winter (December –February)	3	6	12	21

Table 2: Life cycle of different castes of *Apis mellifera* (in days)

Categories Caste	**Duration of stage (Days)**			
	Egg	**Larva**	**Pupa**	**Total**
Queen	3	5	8	16
Worker	3	6	12	21
Drone	3	7	14	24
Life cycle of *Apis mellifera* during different seasons (in days)				
Seasons	Egg	Larva	Pupa	Total
Spring (March-May)	3	5	11	19
Summer (June –August)	3	6	12	21
Autumn (September-November)	3	5	11	19
Winter (December –February)	3	6	11	20

The Egg

The white, sub-cylindrical honey bee egg has one end that is somewhat widened. It weighs about 0.1 gram and is roughly 1.56 mm long. The queen lays every egg in a typical colony. The egg must be neat and cleaned when the queen examines the comb cell prior to egg deposition. The queen inserts her abdomen into the cell and deposits an egg, which sticks to the bottom of the cell by its smaller end, if the cell looks adequate. A thin protective shell called the chorion surrounds the egg and resembles the ovary's follicular cells. The surface of the egg appears to be divided into pentagonal cells under a microscope. The true egg wall, or vitelline membrane, is a thin, non-cellular tissue found inside the chorion. The yolk, the nucleus, and the cytoplasm are all located within the vitelline membrane. In a little mass of cytoplasm, the nucleus is located close to the anterior, or bigger, end of the egg. The remaining cytoplasm creates a network inside the yolk, and outside the yolk in contact with the vitelline membrane is a thinner layer termed as the cortical layer.

Number of Eggs Laid by a Queen

The temperature of the hive, the kind and amount of food the queen is fed, the number of nurse bees in the colony, the availability of fresh nectar and pollen, and the honey and pollen stores in the hive are just a few of the environmental factors that have a direct impact on the queen's ability to produce eggs. The individual egg cells begin their development in the ovarioles' journeys, and after two to three days, they arrive at the oviduct, where they are prepared for laying. The quantity and quality of royal jelly provided to the queen by the nurse bees, as well as the number of ovarioles in each ovary, will determine the number produced.

When fresh nectar or pollen is not readily accessible, a characteristic queen will typically begin laying a few eggs in the early spring. As the weather and food supply improve, she will produce more eggs each day, reaching a maximum of 1,500 to 2,000 eggs each day. It is clear that the queen has to eat more royal jelly during this high egg laying period because the weight of the eggs she lays occasionally exceeds her body weight.

Deposition of the Egg

The bee sting is often acknowledged as a modified ovipositor. Its primary function is to act as weapon by injecting poison into a foe. The curvature of the queen's sting is decurved, and it resembles the curvature of the egg quite closely. During the egg-laying process, the sting of the queen and the sting palpi emerge from a groove, with the sting's curve forming the groove's upper part. Small end of the egg is kept in place temporarily against the cell floor as it travels through this groove with its small end down first. The small end of the egg has a glue like material and adheres it tightly to the cell bottom in an upright position. Probably because their abdomens are too small to reach the bottom of the cells, laying workers usually drop their eggs into the cells. The straight sting of the worker may make it more difficult to place the eggs in the bottom of the cell in an upright position.

General Development of the Egg

Owing to its parthenogentically developed state, the drone egg exhibits some degree of specialization. Sperm are expelled from the spermatheca as the worker egg passes through the duct's entrance before to deposition. One or more of them pass through the micropyle and into the egg. Before the nucleus joins the sperm, it has already undergone modifications. Following fertilization, the merger of the male and female pronuclei forms the segmentation nucleus, which then starts to split. A tiny quantity of cytoplasm envelops each of the cells (referred to as cleavage cells) that arise from the segmentation nucleus's division. The yolk is filled with these cleavage cells. In order to construct the blastoderm, a continuous cellular layer, the majority of them travel to the egg's periphery. The yolk still contains certain cells that aid in its absorption. The thickness of the blastoderm is first almost consistent. The blastoderm's ventral and lateral sections see rapid cell elongation. Soon after, the blastoderm's ventral surface develops two folds. Thus, the blastoderm's mid-ventral portion sinks into the yolk and gets covered by the lateral sections. Consequently, an embryo that appears to have two layers forms. The future ectodermal organs are formed by the ventral and lateral parts of the outer layer.

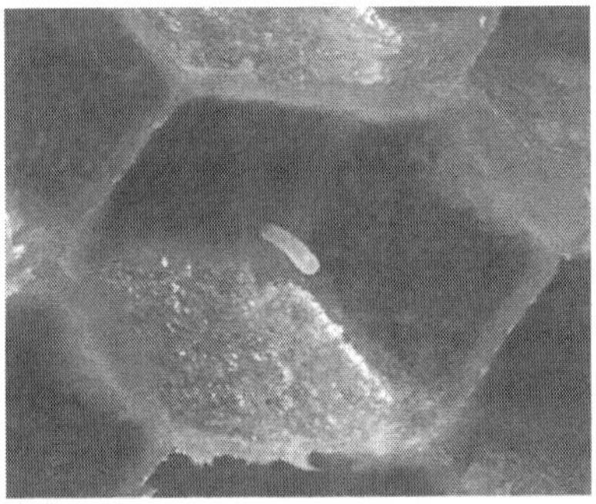

Egg

Larvae

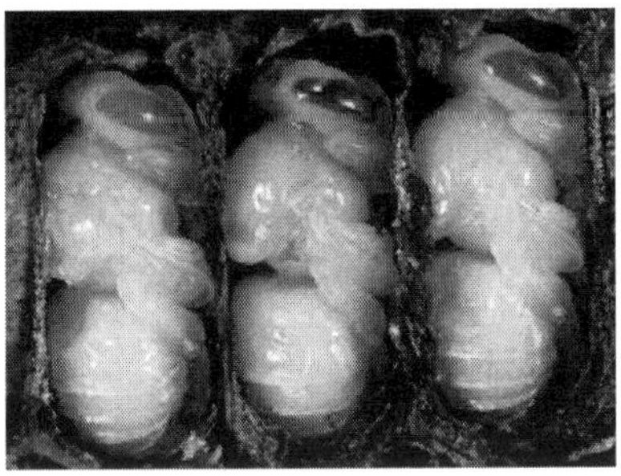

Pupae

Adult

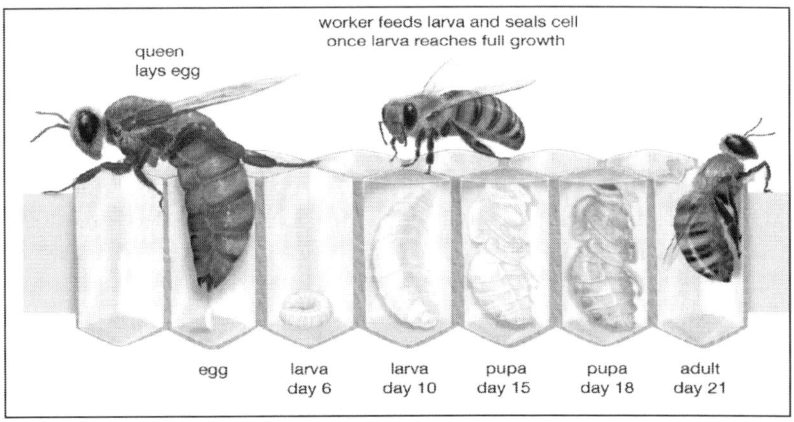

Honey Bee Biology

The serosa, a protective envelope, is formed by the dorso-lateral region. Furthermore, a small number of cells in the dorsal region combine to produce the so-called dorsal organ, which finally degenerates in the yolk and whose purpose is unclear. The organs of mesodermal origin will develop from the section of the blastoderm, or "mesoderm," that submerged into the yolk. The mesoderm forms a single layer along the midline but divides soon after to form inner and outer layers laterally. It's fairly contentious whether an insect embryo contains a real endoderm or not. The development of distinct layers in honey bees that correspond to endoderm in higher forms proceeds as follows: "Blastoderm cells migrate inward to form the rudiments of the mid-gut epithelium, resulting in the production of a discoid swelling at both ends of the middle plate, but outside its bounds." With the exception of the anterior rudiment, where a tiny circular region stays exposed and eventually becomes the floor of the stomodaeal invagination, these rudiments are later covered by the ectoderm. The embryo is clearly segmented after the blastoderm divides into ectodermal, mesodermal, and endodermal parts as the serosa grows. There are 21 distinct portions in all. The mandibular, three thoracic, and antennal divisions all have visible appendages, but lacks in abdomen . A total of 21 segments can be distinguished. Appendages are evident on the antennal, mandibular, and three thoracic divisions but are lacking on the abdomen.

The Larva

The color, shape, and size of honey bee larvae vary greatly . It weighs roughly 0.1 mg and measures 1.6 mm when it first emerges. Its head is not severely sclerotized, its appendages are not as noticeable as they are in many insect larvae, and it is subcylindrical and apodous. The body emerges as head with thirteen circular segments. The freshly emerged larva is white with a faint hint of yellow. The larva develops a fusiform shape with a bigger posterior end as it gets closer to adulthood. Although many beekeepers describe the larvae as milky white, the color eventually turns ivory white. As the larva grows, its bodily divisions become more noticeable. Subconical in shape, the larva's head retracts into the prothoracic segment.

A medial cleft appears to split the head vertically into two subequal lobes when viewed from the frontal aspect. The head is divided into two sections, the lower making about one third and the top approximately two thirds, by a horizontal split. The mandibles, a median flap-like structure, two disc-like swellings, the antennal rudiments, and the united clypeus and labrum are all located in the upper region. The maxillae and labium are located in the lowest portion of the head. The palpi of the adult insect are most likely similar to the spine-like structures seen on the maxillae and labium. Furthermore, the common orifice of the silk glands is located in the labium. Although not in diameter, the thoracic segments are subequal in length. Two pairs of spiracles are found in the thoracic regions, and faint wing rudiments can be seen on the ventro-lateral edges of segments two and three. All abdominal segments are roughly the same size, with the exception of the final two, which are smaller. The anal aperture is situated on a tiny lobe that is present on the ultimate division.

The first eight segments of the abdomen have spiracles. The lateral and ventral regions of the abdomen are divided by a deep groove that runs from abdominal segments one through eight. The lateral borders of these identical segments give rise to swollen regions known as the epipleural lobes. A worker bee can only weigh 159 mg attained between 114 and 122 hours.

The Pupa

The final larval stage, described as the prepupal stage, is when specific changes start to appear. The larva spins its cocoon and goes into quiescence. Within the larval head capsule, the head shrinks and the mouthparts elongate. The growth of the complex eyes and antennae starts. The abdominal and thoracic appendages show through. The insect grows quickly, and if it is taken out of its larval skin before its last moult, it resembles an adult very much.

The pupa appear fully grown after the sixth molt. The following are some notable modifications from the larval stage: 1. the head and appendages are well developed.2. The head is angled ventrally at this point. 3. There has been a significant growth in the mesothorax. 4. The thorax and the first abdominal segment have united. 5. There is a profound constriction dividing the remaining portion of the abdomen from the rest of the body. The pupa's coloration varies as it gets older.

The compound eye is where they are most noticeable, as they turn pink, then purple, and then black. The abdomen and thorax eventually turn brown and acquire the color of an adult bee. The development of branching hairs and other integumentary processes are examples of further changes. For worker and drone bees, the pupal stage lasts eight to nine days, while for the queen, it is just five.

Development of the Larval and Pupal Stages

A worker bee deposits just a small amount of royal jelly, a food source for larvae, within the egg cell just prior to hatching. It is believed that the royal jelly's moisture is necessary for the egg to hatch. As a result of "mass feeding," the nursing bees abundantly feed the larva with royal jelly so soon after hatching that it is literally floating in the liquid. After three days, a new food is given out that contains pollen. Furthermore, feeding practices are always evolving.

The larvae receive nourishment ostensibly when they require food, rather than always having an abundance of food. It is commonly acknowledged that a larva's development into a worker or a queen depends on the kind of food it consumes. It has been claimed that during the entire feeding period, larvae that will eventually become queens are fed royal jelly.

However, the primary cause of either the worker or the queen's production is the differing amounts of vital nutrients that the worker and queen larvae ingest. worker bees are malnourished. The quantity and quality that worker larvae receive is not only lower than that of queen larvae, but also lower. It appears that as worker larvae develop, these qualitative changes can be observed in the decline of specific vitamins. One extremely labile component that exists in royal jelly is essential for queen differentitation.

B. Honey Bee Colony

Queen Bee

All other bees in the colony are descended from the queen bee. Its tiny wings and long abdomen help to identify it. The queen's job is to deposit eggs. By virtue of its pheromones, the queen sustains the colony alive. The volume of food the workers bring in and the amount of space available for brooding inside the colony determine how productive she is.

Honey Bee Colony

A queen bee's daily egg production exceeds 1500. She will lay up to 2,500 eggs each day if it is honey flow season and there are enough cells available. Because of its structural similarity to a cow's teat, the queen emerges from the queen cell, which is located at the base of the comb and resembles a little cup. In India, this cell is commonly referred to as the teat of a cow. Following her emergence, the newly formed queen battles any other queens she comes across and destroys the colony's remaining queen cells.

Usually, the virgin queen stays in the colony for a few days to feed, grow stronger, and allow her reproductive organs to reach a slightly more mature stage. The queen will depart the colony after 6–8 days to embark on her nuptial flight, which takes place 30 meters above the earth. There, she will mate with numerous powerful drones who can soar alongside her since she is a superior pilot than drones. The queen goes back to the hive after mating so as to lay the eggs. A queen can live up to five years on average, but she can only lay eggs for two of those years.

The Worker

A colony has thousands of workers who carry out various distinct tasks within the colony, such as cleaning, guarding, foraging, and raising the young. Compared to the drones and queen, they are smaller. A. florea colony has roughly 8000–25,000 workers, A. mellifera colony has 40,000–50,000 workers , A. cerana has 20,000–40,000 workers, and A. dorsata colony has 50,000–80,000 workers. Worker bees use their modified ovipositor sting to defend their colonies, and when they sting, they release venom. Under some circumstances, workers may lay eggs that, because they never mate and lack sperm to fertilize their eggs, turn into drones. On the other hand, worker management takes place and workers eat the eggs laid by other workers in a typical queen right colony. Unlike A. mellifera, A. cerana can have a comparatively large number of workers that lay eggs. When a worker is young, they work indoors. When they are old enough, usually after 21 days, they are granted liberty to go outside foraging.

The Drone

Drone bees are effortlessly recognized by their dark color and the eye that touches the top of their head. Their sole purpose is to eat the food within the hive and fertilize the queen. They have no stingers, thus they don't sting. As "haploid" creatures, drones only have half the gene pairs that the "diploid" workers and queen have. There are roughly hundreds of drones in a colony; drone cells have larger cappings than worker cells, and commercial beekeepers in India decap these drone cells when the drones eat the honey that has been stored. They have about 75–80% more facets in their compund eyes than the workers or queen, which gives them superior navigational skills over the other two castes.

Caste Distinction of Honey Bees

Characters	Queen	Drone	Worker
Number in a colony	One	About 100	50, 000-60, 000
Size	Largest	Medium	Small
Sex	Fertile female	Fertile male	Unfertile female
Life span	4-5 years	Dies after mating	4-6 weeks
Head	Round	globular	Loner and more triangular than queen
Antennae	Relatively longer but with same no. of segments	Longer than queen and worker and composed of one segment more (13)	-
Compound eyes	Smaller and located far apart	Meet on the top of the head	Well developed and situated far apart on the sides of the head
Mandibles	Well developed and notched	Chisel like and notched	Smooth and rounded
Tongue	Normal	Normal	Long and well developed
Thorax	and deeper than worker`	Normal	Width of thorax is less than queen and drone
Legs	Normal	Normal	The three pairs of legs are modified are modified for 1) Antennal cleaning 2) Wax picking _ 3) Pollen collection
Wings	Appear shorter than worker because of longer abdomen	Normal	Normal
Wax plates	Absent	Absent	present
Abdomen sting	Longer and triangular	Rectangular with blunt end absent	Triangular and striped barbed sting

CHAPTER 5
Honey Bee Foraging

India is a gigantic country with a variety of climate zones that offer rich vegetation to honeybees. Through their foraging activities, honeybees gather pollen and nectar, which together provide them with the protein and carbohydrates they need to meet their nutritional needs. Only the forager bees will carry out foraging since honeybees have a division of labor. Nectar and pollen collectors are the two categories of foraging bees. They must visit hundreds of flowers to gather the entire load of nectar or pollen. On top of this, the foragers collect water and plant resins termed propolis in the case of stingless and Italian bees.

Over 70% of the world's produced crops are reportedly pollinated by bees, which are the most efficient pollinators of both crops and natural flora. Additionally, it has been stated that domestic bees pollinate approximately 15% of the 100 major crops. These are some of the crops with rich sources of pollen and nectar that bees benefited from, as honeybees in India had a wide variety of flowers to choose from.

Rich nectar sources include *Eucalyptus sp., Syzygium cumini, Anacardium occidentale, Nephelium litchi, Azadirachta indica, Callistemon citrinus, Glycine max, Cajanus cajan, Hevea brasiliensis, Acacia catechu, and Dalbergia sissoo.* Amongst the abundant sources of pollen are *Zea mays, Psidium guajava, Sesamum indicum, Sorghum bicolor, and Helianthus annuus.* Some of the plants that give bees a lot of pollen and nectar are *Citrus sinensis, Coriandrum sativum, Cucumis melo, Eucalyptus spp., Musa spp., and Pongamia pinnata.* Bees get their food from *Carica papaya, Cocos nucifera,* and *Musa spp.* all year round.

Honey bees have a mutualistic relationship with flowers in which the bees provide nectar and pollen in exchange for the flowers' assistance in pollination. The foraging activity of bees determines their survival efficiency and is positively connected with colony size. Diverse circumstances within and beyond the colony incentivize worker bees to engage in foraging. The age of the workers, the existence of a queen, the strength of the colony, the demand for food and storage, pests and diseases, etc. are examples of internal colony factors. The ambient weather, which comprises, among other things, temperature, precipitation, relative humidity, wind speed, daylight hours, and the presence of forage flowers, is an external factor. When honey bees begin foraging, they are exposed to a wide range of ambient weather conditions, which can affect their productivity and cause them to transform. Each climatic component affects honey bee foraging behavior differently.

CASHEW

CORIANDER

COCONUT

GOAT WEED

NIGER

SUNFLOWER

RAPESEED

RADISH

MUSTARD

TAMARIND

CUCUMBER

SUN HEMP

Foraging by Bees

1. **Sunflower:** It is a cross-pollinated crop. The pollen of the plant cannot fertilize ovary of same plant. Pollen source should be from different plant. Hence, honey bees acts as important agents for pollination in sunflower. In sunflower, yield increases even up to 600 per cent due to bee pollination. It improves quality and quantity of seeds. Oil content also increases by 6.5 per cent in seeds. To achieve this it requires five strong C. *indica* colonies or three A. mellifera colonies. Mostly irrigated crop is preferred by bees.
2. **Cucurbitaceous vegetables:** Cucurbits are monoecious with staminate and pistillate flowers in same plant. Due to bee pollination fruit set increases up to 30 to 100 per cent.
3. **Alfalfa or lucerne:** These plants have tubular flowers with 5 petals joined at base. They posses one large standard petal, 2 smaller petals on sides and 2 keel petals holding staminal column. When bee sits on a keel petal, stamina column strikes against standard petal resulting in shattering of pollen. This is called *tripping*. Seed set occurs only if bee sits to trips the flowers.
4. **Coriander:** In coriander yield increases up to 187 per cent due to pollination.
5. **Cardamom:** It is an important commercial crop depending on bees for pollination. Here yield increases up to 21 to 37 per cent.
6. **Gingelly:** Another oilseed crop where bee pollination causes 25 per cent increase in yield.
7. **Fruit Trees:** All kinds of apple citrus, litchi, peach, apple, guava, *jamun*, date palm, apricot, quince, pear, almond, plum, loquat, phalsa, and cashew.

Cultivated field crops: Pigeon pea, lentils, clovers, Lucerne, mustard, rape, linseed, sesame, gingelly, buck-wheat, Cambodia, safflower, millet and sunflower

Vegetables: All cucurbitaceous plants, okra, beans, turnip, radish, onion, brinjal, and sweet potato.

Timber trees: Neem, *Cassia fistula*, Acacia, *Albizzia* spp., Kachnar *(Bauhinia purpurea)*, eucalyptus, sandle-wood, raintree, wild cherry.

Natural and Ornamental Flowers

Cosmos, shoe flower, Golden rod, Cup & saucer, *Tecoma stans*, zinnia, coral creeper *(Antigonon leptopus)*, rose, rangoon creeper, aster, wild rose *(kuja)*, hydrangea, violet, portulaca, poinsettia, honey suckle, corn flower, coreopsis, dandelion etc.

WHITE SWEET CLOVER

KARANJ

CITRUS

DRUMSTICK

EUCALYPTUS

SESAME

LITCHI

SOAPNUT

BRINJAL

WATERMELON

ONION

SWEET POTATO

Cosmos

Hibiscus

Golden rod

Aster

Tecoma stans

Zinnia

Edward rose

Coral creeper

Rangoon creeper

CHAPTER 6

Honey Bee Communications

Honey bees use all of their senses to find the best flowers including: smell, color, shape, location, petal textures, and time of day. But what does a bee do when she wants to tell her sisters (other worker bees) what she has discovered?

Bee learning and communication includes cognitive and sensory processes in all kinds of bees, that is the insects in the seven families making up the clade Anthophila. Some species have been studied more extensively than others, in particular *Apis mellifera*, or European honey bee. Color learning has also been studied in bumblebees.

Honey bees are sensitive to odors (including pheromones), tastes, and colors, including ultraviolet. They can demonstrate capabilities such as color discrimination through classical and operant conditioning and retain this information for several days at least; they communicate the location and nature of sources of food; they adjust their foraging to the times at which food is available; they may even form cognitive maps of their surroundings. They also communicate with each other by means of a "waggle dance" and in other ways.

Associative learning is an ability that honey bees specialize at, and numerous components of operant and classical conditioning have similarities in honey bee behavior to that of vertebrates. Foraging successfully demands this kind of learning. For instance, if a plant offers little in the form of rewards, honey bees will not return often. Until the plants stop producing nectar or the weather changes, a lone forager will visit various flowers in the morning and, if there is enough reward in a particular kind of flower, will visit that type of flower for the greater part of the day.

Color Learning in Honeybees

Numerous studies have shown that *Apis mellifera* is capable of distinguishing between colors and has memory. Scientists Karl von Frisch and later Randolf Menzel started investigating several facets of color learning and color vision in bees in the early 1900s.

Color Discrimination

In 1919, Austrian naturalist Karl von Frisch raised the concern, "Do bees have color vision?" which sparked investigations into the question of color vision in honey bees. He carried out an elaborate experiment that defined the bees' capacity for associative learning in addition to their ability to distinguish between colors. Initially, he trained his bees to feed from a small bowl that contained sugar water that tasted like nectar.

To make the color of the dish noticeable to the bees as they arrived and fed, this dish was set on a piece of blue-colored cardboard. Von Frisch then surrounded the blue piece with identically sized pieces of cardboard in various hues of gray, each with a dish. The bees should visit one or more of the gray pieces as frequently as the blue piece because they are colorblind, but he discovered that most of the bees headed straight to the blue piece of cardboard where they had previously collected their reward. The gray pieces that had not been awarded were mostly disregarded by the bees. Von Frisch observed the same conclusions when he repeated the experiment with different colors, such as violet and yellow. subsequently, multiple scientists tested vertebrates' ability to see color using this experimental setup.

In 1973, Karl von Frisch was awarded the Nobel Prize for his breakthrough in deciphering bee language. He placed a honey bee hive in a field with a single nectar feeder for the bees to gather from in some of his early trials. After that, Dr. Frisch sat and observed the bees in their hive. He observed the bees for several hours before realizing they were wriggling inside the hive. Throughout the day, he measured and tracked animals as they engaged in this behavior. Dr. Frisch saw that by doing this, the bees were dancing at an angle that matched the speed at which the sun was rising in the sky. He found out that the bees were directing their sisters toward the sugar feeder in relation to the sun through dance. Although Dr. Frisch and his colleagues subsequently translated more elements of the bee dance language that represented distance, there is still a great deal we do not know about these amazing tiny insects.

Honeybee communication is essential to bee existence. Honeybees generally communicate in two ways.

1. The pheromone activity
2. Dance (waggle dance and round dance)

The Pheromone Activity

Bees produce chemicals called pheromones, which they employ to communicate with other members of their colony. Pheromones come in a variety of forms and can cause a range of reactions. The first is an alarm pheromone, which bees employ to warn others of impending danger. Additionally, queen bees create unique pheromones that aid in population control. She can motivate the remainder of the colony to continue producing for the hive, for instance, if she produces a pheromone that alerts the others to her presence and health. In order to draw drones, Queen produces a substance known as a pheromone. The chemicals that bees emit aid in

1. Supporting the community in identifying the queen.
2. Lowering the quantity of eggs that working bees lay.

3. To maintain bee unity during swarming in order to establish a new colony.
4. For interacting with other hive bees.
5. Spreading their wings to help the hive cool.
6. To remain near swarming.

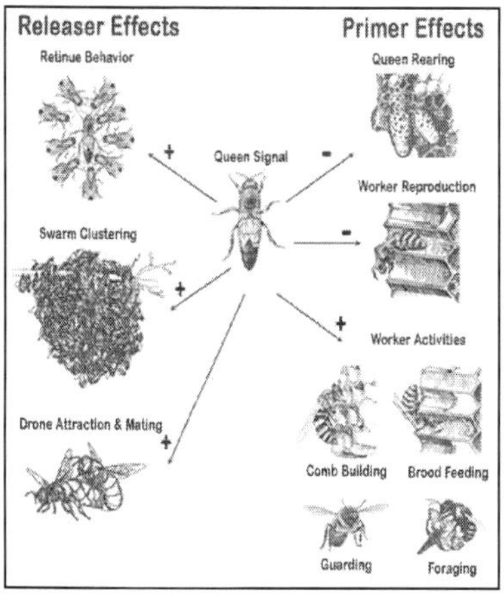

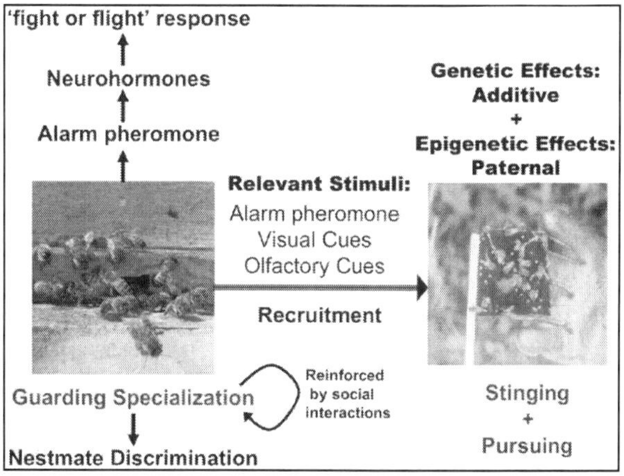

Any chemical signal that individuals of a species use to communicate with one another is called a pheromone. The most extensive research on pheromones has been conducted on insects, although it is possible that more complex species also have pheromones. There is insufficient evidence to verify their presence in humans. More complicated species may only preserve vestiges of pheromone-emitting organs since their individuals have evolved more advanced communication strategies.

A molecule known as a pheromone causes a natural behavioral reaction in another individual of the same species. There are numerous pheromones that influence physiology or behavior, including sex, alarm, and food trail pheromones. Though many plants and vertebrates also employ pheromones for communication, their function in insects has been extensively studied. Bees employ the pheromone isopentyl acetate as a physical illustration of an alert system. Guard bees have the ability to raise their abdomen and release pheromones, which they can then spread by beating their wings. High amounts of pheromones are deposited along with the venom in this fashion, alerting the entire hive of bees to an impending threat and motivating other bees to sting in the same location.

Ants navigate to and from the nest and food sources using pheromone trails. When a pheromone-free object, like chalk, blocks the trail, the ants will either stop and wander aimlessly or turn around. Pheromones were a reliable means of signaling and navigation for insects before the invention of eyes.

Based on the Greek words pherein, which means "to transport," and hormone, which means "to stimulate," Peter Karlson and Martin Luscher coined the term "pheromone" in 1959. Not long after Butenandt identified the first such chemical, Bombykol (a chemically well characterized pheromone secreted by the female silkworm to attract mates), they introduced the name to represent chemical cues from conspecifics which activate inherent behaviors.

Pheromones emitted by individual bees into the hive or surroundings are chemical mixes known as honey bee pheromones, which alter the physiology and behavior of other bees. Because they have 15 recognized glands that create a wide variety of substances, honeybees have one of the most complicated pheromonal communication systems in the natural world. Pheromones are created in liquid form and are transferred in liquid or vapor form by direct contact. Both volatile and non-volatile pheromones exist.

Pheromones are chemical transmitters that cause other bees to react. They are secreted by a queen, drone, worker bee, or laying worker bee. The antenna and other body parts of the bee receive the chemical information. There are two types of pheromones found in honey bees (*Apis mellifera*): primer pheromones, which have long-term effects, and releaser pheromones, which have short-term effects.

The recipient's physiology is altered by primer pheromones. Releaser pheromones alter the recipient's behavior. Releaser pheromones are transient in nature and cause the receiving bee to react behaviorally almost instantly. A pheromone has the ability to function as a primer and releaser under specific circumstances. Pheromones can be simple compounds or complicated mixtures containing many distinct substances in varying proportions.

Types of honeybee pheromones: There are mainly two types of pheromones present in honeybees.

1. **Primer pheromones:** Bee development events alter in response to primer pheromones. It works by using gustatory sensilla.
2. **Releaser pheromones:** Certain species have the ability to lure potential mates up to two miles away using strong chemicals called releaser pheromones. Though it swiftly degrades, this kind of pheromone typically triggers a speedy response. A primer pheromone, on the other hand, starts and lasts longer.

Releaser pheromones again divided into following types.

1. **Aggregation pheromones:** Produced by one or the other sex, aggregation pheromones attract individuals of both sexes of honeybees.
2. **Alarm pheromones:** Some species release a volatile substance when attacked by a predator that can trigger flight (in aphids) or aggression (in bees) in members of the same species. Pheromones also exist in plants: certain plants emit alarm pheromones when grazed upon, resulting in tannin production in neighboring plants. These tannins make the plants less appetizing for the herbivore. E.g. 2-Heptanone as a putative alarm pheromone in rats.
3. **Epideictic pheromones:** Recognized in insects, epideictic pheromones are different from territory pheromones. According to Fabre (translated

from French), " Females who lay their eggs in these fruits deposit these mysterious substances in the vicinity of their clutch to signal to other females of the same species so that will clutch elsewhere".

4. **Territorial Pheromone:** Laid down in the environment, territorial pheromones mark the boundaries of an organisms territory.
5. **Trail Pheromones:** In honeybees, trail pheromones are often present. Honeybees, for instance, use these non-volatile hydrocarbon pheromones to indicate their travel routes. As some honeybees carry food back to the nest, they leave a first pheromone trail. This path works as a guide and draws in more honeybees. Because it evaporates quickly, the pheromone trail will be continuously recreated as long as the food supply is around. The trail marking stops when the supply starts to run low.
6. **Sex pheromones:** Sex pheromones in animals signal a female's availability for reproduction. Additionally, pheromones that communicate information about an animal's genotype and species can be released by males. In order to attract a partner, many insect species emit sex pheromones, and many lepidopterans, such as moths and butterflies, can identify a possible mate up to 10 kilometers away. Gametes can employ pheromones to follow the gametes of the opposing sex in order to fertilize. Pheromones are also utilized to identify sow oestrus.

Other Pheromones

This classification, based on the effects on behavior, remains artificial. Pheromones fill many additional functions.

1. Nasonov pheromones (Worker bees) released by nasonov glands.
2. Royal pheromones (Bees).
3. Calming (appeasement) pheromones (mammals).

Types of Queen Honeybee Pheromones

Queen Mandibular Pheromone (QMP)

One of the most significant sets of pheromones in the bee hive is QMP, which is released by the queen. It has an impact on worker bee social behavior, hive maintenance, swarming, mating behavior, and suppression of ovarian growth. There may be immediate or long-term consequences. QMP contains a variety of substances, including aromatic compounds and carboxylic acids. The importance of the following substances in attracting workers to their queen in a retinue and for other purposes has been demonstrated.

1. (E)-9-oxodec-2-enoic acid (9-ODA)-inhibit queen rearing as well as ovarian development in worker bees; strong sexual attractant for drones when on a nuptial flight; critical to worker recognition of the presence of a queen in the hive.

2. (R, E)-(-)-9-hydroxy-2-enoic acid (9-HDA) promotes stability of a swarm, or a calming influence
3. (S, E)-(+)-9-HAD
4. Methyl-p-hydroxybenzoate (HOB)
5. 4-hydroxy-3-methoxy phenylethanol (HVA)

Queen Retinue Pheromone (QRP)

Additionally, the following substances have been detected; only coniferyl alcohol is present in the mandibular glands. The queen retinue pheromone (QRP) is the concoction of the four chemicals below and the five QMP components. These nine substances are critical to the worker bees' entourage surrounding their queen.

1. Methyl (Z)-octadec-9-enoate (methyl oleate)
2. (E)-3-(4-hydroxy-3-methoxyphenyl)-prop-2-en-1-ol (Coniferyl alcohol)
3. Hexadecane-1-ol
4. (Z9,Z12,Z15)-octadeca-9,12,15-trienic acid (linolenic acid)

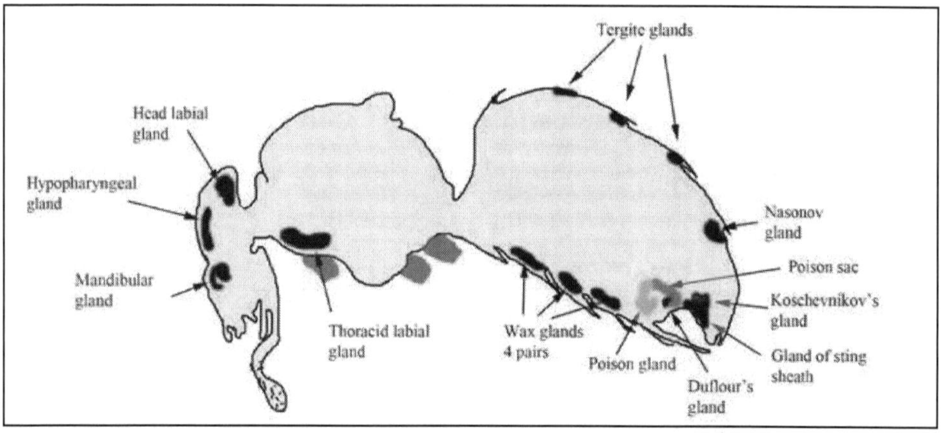

Fig. 1: Honey bee Pheromonal glands

Dance Communication

Two fundamental food dances are done by honey bees. The first dance, referred to as a "round" dance, is focused on food sources that are within 50 meters of the hive. To signal to the other worker bees where the food is, the honey bee will make a series of little circles. Honey bees "waggle" in a figure-eight pattern to locate food sources farther from the hive. It does this by buzzing and vibrating its wings; the intensity of this signalizes the distance to the food source. In order to aim toward the food supply, the bee also angles its body with respect to the sun.

The honeybee's dance language is a form of social behavior. It is utilized by a single worker to convey two or more pieces of information to one or more other workers (the location's direction and distance from the food source, which is typically a flower's path). In 1919, Karl Von Frisch, an Austrian who lived from 1886 to 1982, started training European honey bees, *Apis mellifera,* to become feeders. In 1973, he was awarded a Nobel Prize for his studies on honeybees. There are two dances which are performed by the honeybees

A. The round dance

B. The waggle dance

A. Round Dance

A forager bee will perform a round dance if the supply of food is lesser than fifty meters from the bee hive. To accomplish this, she runs in little circles before abruptly turning back to her initial path. She has the option to do the dance multiple times at the same spot or at a different location. Upon the conclusion of the circular dance, she frequently gives food to the bees trailing behind her. A round dance expresses simply the distance and not the direction.

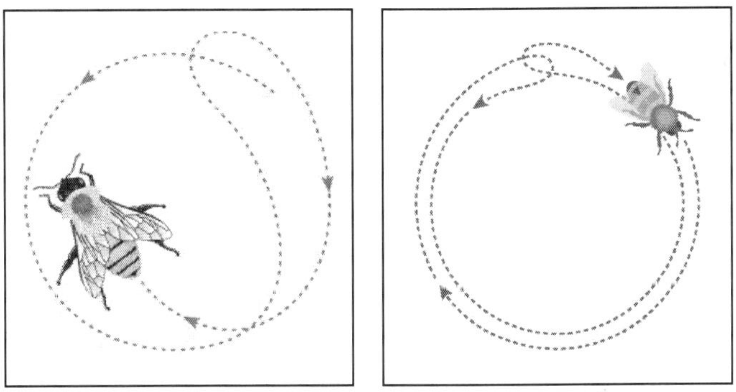

B. Waggle Dance

Bees execute the Waggle Dance, also known as the Wagtail Dance, when they are foraging at food sources more than 150 meters from the hive. The other worker bees can tell direction and distance from each other through this dance.

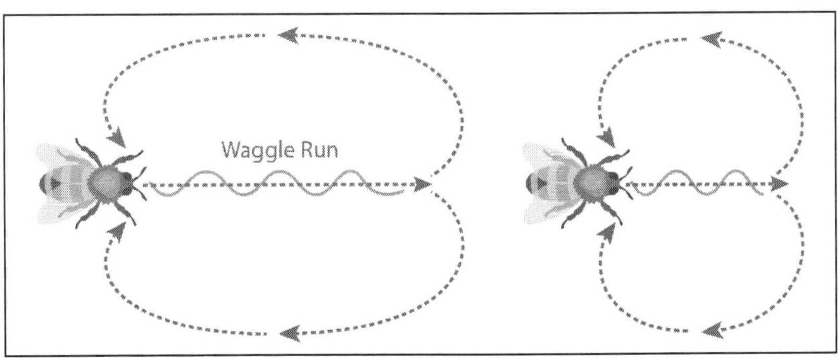

Distance

A waggle dancing bee runs a short distance straight ahead; circles back to the starting place, run straight ahead again, and then make a semicircle in the other direction to complete a full figure eight circuit. The bee's body, particularly its belly, vigorously wags sideways while it runs the straight-line course of the dance. This vibration provides the tail its wagging motion. The length of the dance's waggling part rises along with the food source's distance. Scout bee hunts for fresh food and dances whereas, recruit bees pursue foraging after dances. During the straight part of the dance, when the bee is waggling, the direction the dancer faces reveals the direction of the food supply. If she waggle with her face pointed straight up, the food supply could be located in the sun's direction.

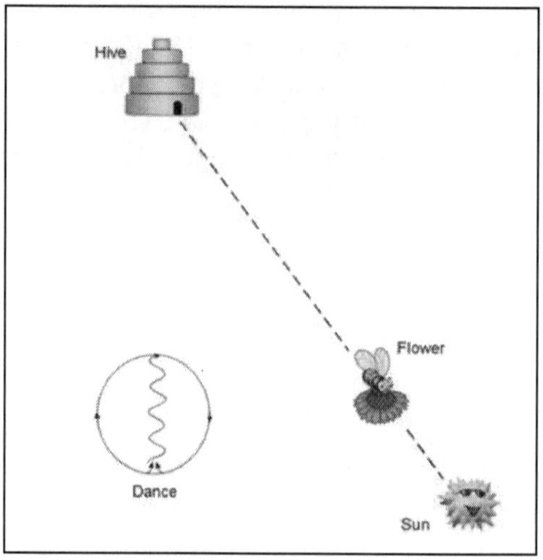

The food supply is 60 to the left of the sun when the worker bee waggles at an angle of 60 to the left of upward.

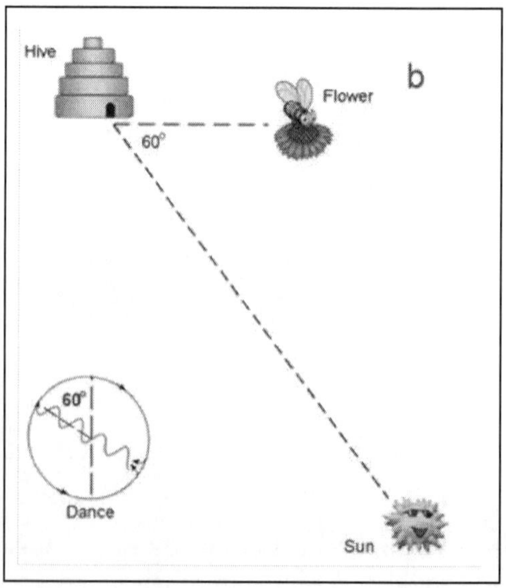

The food supply may be located 210 to the right of the sun when the worker bee waggles 120 to the right of upward. During the waggle run, the dancer makes sounds that aid the newcomers in determining their path in the hive's darkness.

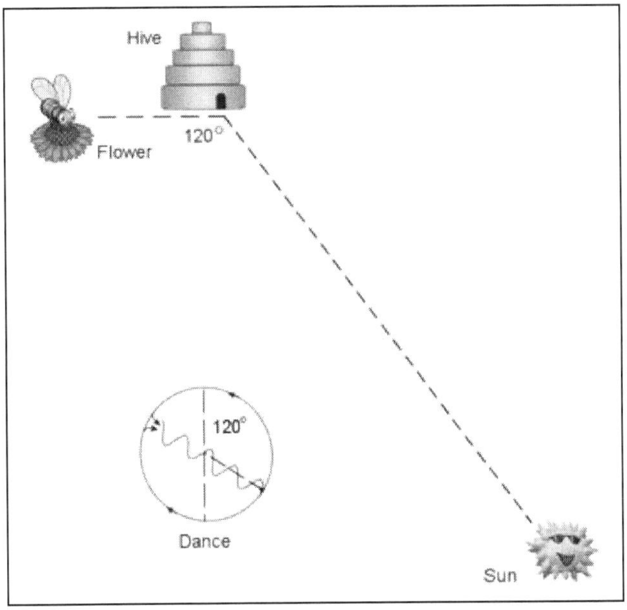

Waggle Dance Says

Direction	Angle of waggle run
Distance	Number of waggles/ run dance tempo (of circuits/unit time
Quality	The duration of the dance (total waggle runs)
Taste and smell	Dancer gives the free samples

CHAPTER 7

Beekeeping Equipment

The equipment utilized for beekeeping has evolved over the years along with the trade. The Langstroth bee space (1851), the Johannes comb foundation (1857), and the honey extraction techniques developed by Frang von Hruschka had a significant impact on beekeeping in India and led to a dramatic change in the practice. Beekeepers in the subcontinent were encouraged to switch to movable frames because they are more user-friendly and because modern beekeeping equipment makes work easier for commercial handlers. Traditionally, beekeepers in India kept bees in baskets, wooden logs, underground hives, clay pots for stingless bees, etc.

Honeybee Hive

A beehive is the place where a single honeybee colony lives and works. It is made up of several components, including an inner and outer cover, a bottom board, a brood chamber, a super chamber, and a hive support. The hive stand is constructed comprised of an iron or wooden pole that is fastened to the ground. It can have four legs made of iron or PVC (polyvinyl chloride) or it can have a single leg made of iron.

Each has benefits and drawbacks. For example, a single stand can be easily attached with an ant pan, but a four-legged stand can be easily moved around because it is not fixed to the ground like a single-legged stand is. The bottom board is detachable from the hive stand and rests on the stand. The brood chamber, which is located above the bottom board, is made up of brood frames, which are likewise made of wood and are arranged vertically and parallel to one another. The brood frames serve as a home for honeybees, where they raise their larvae in the comb that is constructed within; it also contains pollen and some honey for the bees' daily consumption.

Typically, a brood box holds a single queen bee. The queen deposits her eggs, one within each comb cell. After hatching into larvae, the eggs become adult bees. The apiculturist installs a super chamber atop the brood chamber once it is densely occupied with bees. Similar to the brood chamber, the super chamber is similarly made up of parallel and vertically placed frames. It is the location of our interest since bees use it to store honey. To stop the queen from entering the super chamber, a queen excluder is positioned between the brood and super chambers.

The beekeeper may choose to attach a second brood chamber to the first before attaching the super chamber if the queen lays a lot of eggs. Because there are more worker bees collecting nectar in these situations, there will be a higher production of honey. The purpose of the honey and pollen kept in the brood chamber is solely for the developing larvae; the beekeeper is not allowed to remove them. The super chamber's standard height is one-fourth that of the brood chamber. Because the bees will be gathering more nectar throughout the honey flow season, there will be more super chambers.

Fixing a second or third super chamber to the first is not necessary if the beekeeper takes small bursts of honey out of the hive. There will be a top cover over the super chamber that serves as the hive's ceiling.

Types of Beehive

The Langstroth hive, used for *A. mellifera*, the BIS hive (Bureau of Indian Standards) for *A. mellifera* and *A. indica*, and the Newton and Marthandam hives for *A. cerana* belong to the various kinds of hives used in India.

Langstroth Hive

Stand: Any four-legged, 15–25 cm-tall support will do. Its upper measurements should be such that the bottom board is suitably supported.

Bottom board: It can be constructed by nailing two wooden boards together with wooden rods, or by taking a piece of wood that is 550 mm long, 406 mm wide, and 22 mm thick. A 550 mm long, 22 mm wide, and 22 mm thick wooden rod is nailed at either end of the longer side, and a second, 363 mm x 22 mm wooden rod is nailed at the back. The entry rod on the front measures 363 mm by 22 mm by 22 mm. Its center entrance is 75 mm long and 22 mm deep. Two wooden blocks measuring 75 mm by 38 mm by 22 mm each should be prepared in case the entrance needs to be reduced.

Brood chamber: A rectangular box with no top or bottom, it is constructed of wood that is 22 mm thick. Its measurements are listed as follows: its length is 500 mm on the outside and 456 mm inside; its breadth is 406 mm outside and 363 mm inside; and its height is 238 mm. Across the breadth of the planks, a rabbet measuring 16 mm in depth and 13 mm in width is carved.

Frame is comprised of a bottom bar, two side bars, and a top bar.

1. Top bar dimensions: 475 mm x 25 mm x 22 mm. It is 25 mm long and is carved to a thickness of 9 mm on both sides. Its lower side features a groove in the center where the comb foundation sheet can be fixed.
2. Side bars: Each is 226 mm long and 9 mm thick in wood. Each has a base half that is 25 mm broad and an upper part that is 34 mm wide. To make room for the top and bottom bars, respectively, each is carved out from the middle section at each end. Each side bar has four holes for wiring the frame.

3. Bottom bar dimensions: 440 mm 19 mm 9 mm. The frame's external dimensions are 440 x 228 mm.
4. To make the frames stand 34 mm apart, set two 15 mm staples into the top bar on the opposite side. All frames ought to be made in either Hoffman or staple-spaced type. The frame should be wired with 28 gauge tinned wire.

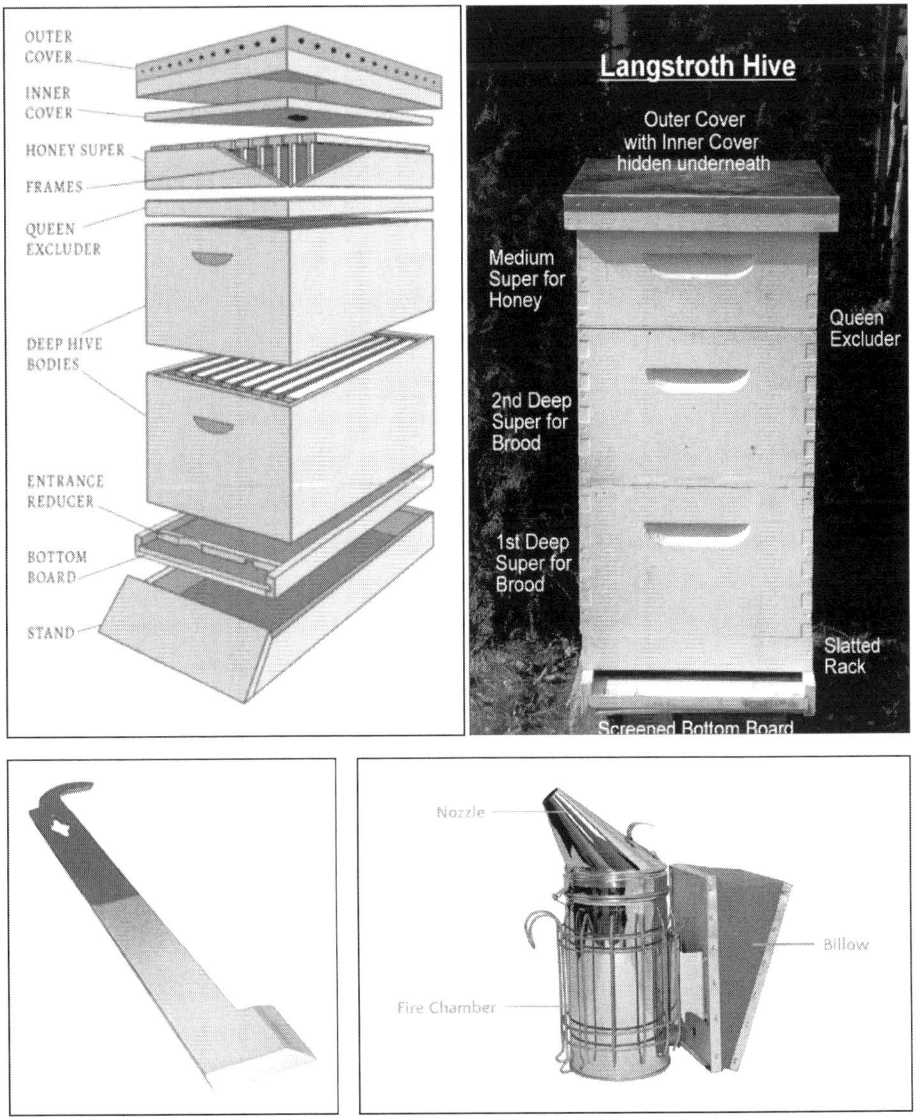

Bee Keeping Equipment

Honey Harvester

Comb Foundation Sheet

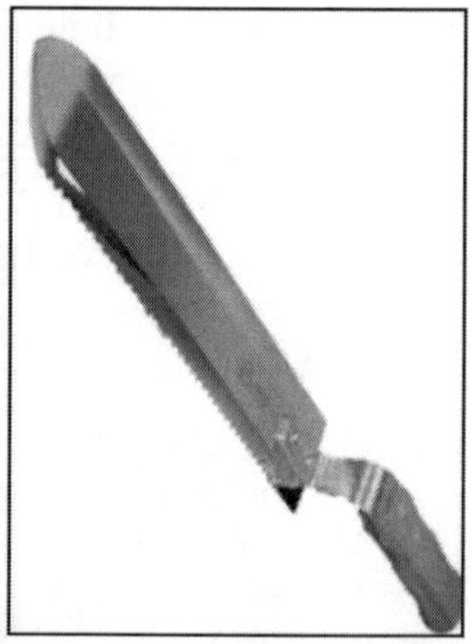

Uncapping Knife

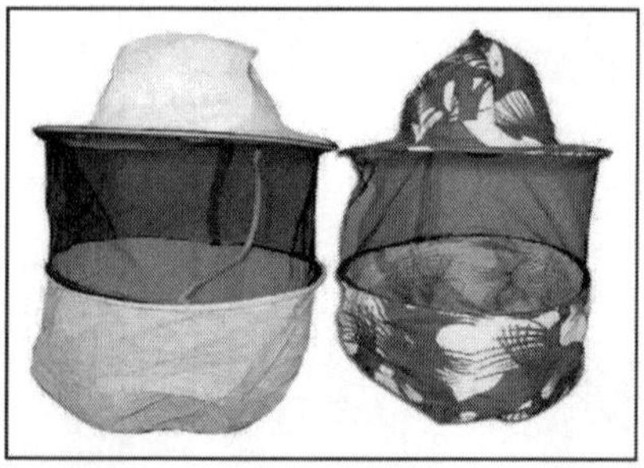

Bee Veil

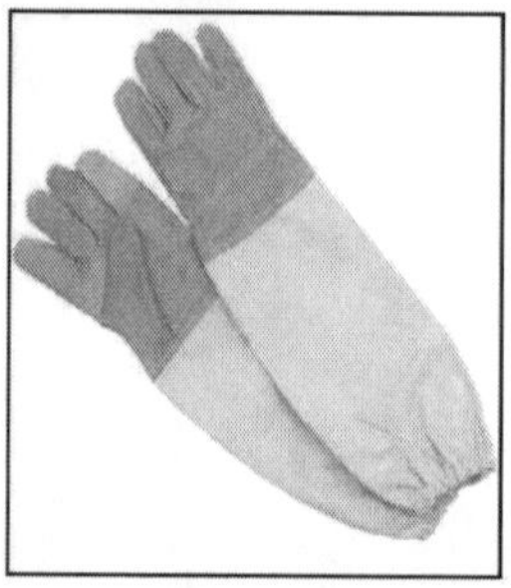

Gloves

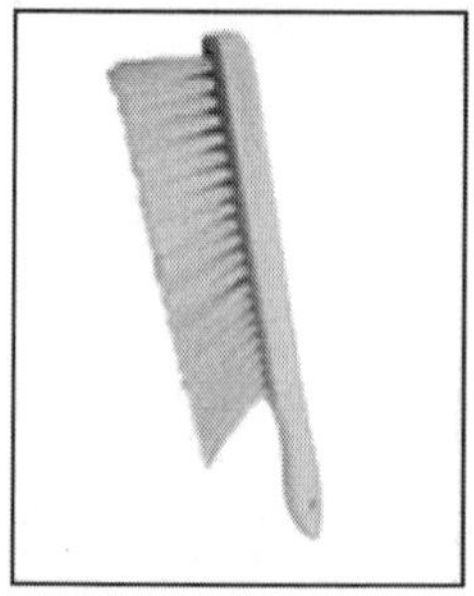

Bee Brush

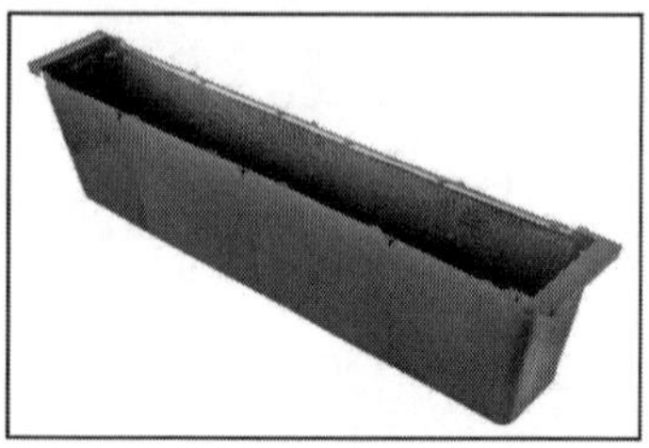

Feeder

Queen Excluder

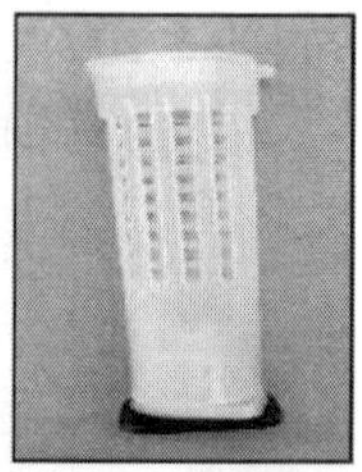

Queen Cage

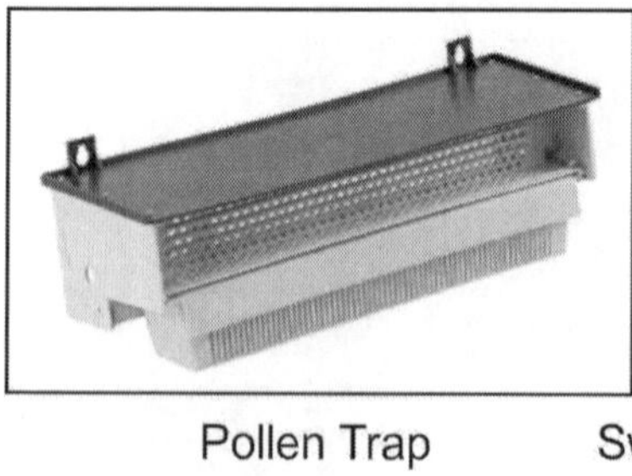

Pollen Trap

Swarm Catching Bag

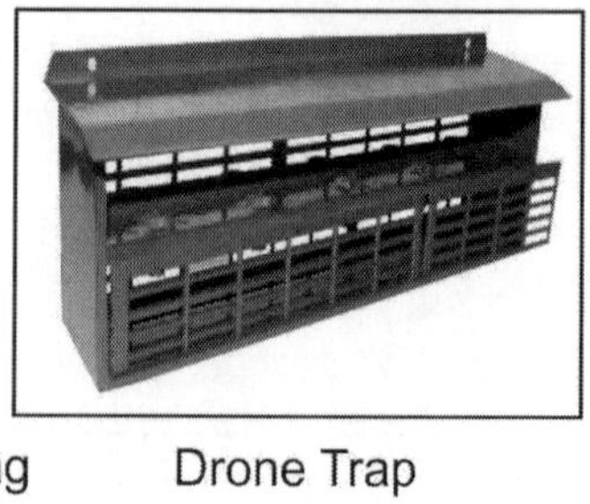

Drone Trap

Bee Keeping Equipment

Super: The brood chamber and brood chamber frames' respective dimensions should match those of the super and super frames, respectively.

Inner cover: This wooden board serves as a cover for the super, or brood chamber, depending on the situation. Its dimensions are 406 mm wide, 500 mm long, and 9 mm thick. It has a hardwood bar that is 22 mm broad and 9 mm thick that is fastened to each of its four sides.

Top cover: To keep rainwater out of it, a metallic sheet is placed over a 9 mm thick hardwood board that is fastened to a rectangular frame that is 50 mm high. Its inside dimensions are 425 x 525 mm. It lies slackly across the hive.

Newton Hive

Stand: A roughly 10-cm-diameter wood log that has been thoroughly soaked in wood preservative is buried far below the surface of the earth. A 40 x 30 cm board is fastened to the top of the 20–30 cm length that is left above the ground using long nails and screws. This platform is where the hive is situated on the log.

Bottom board: This plank has beadings on three sides, which the hive body slides snugly into. It is somewhat longer and broader than the brood chamber by 25 mm. The alighting board is the expansion of the front.

Brood chamber: It's a box without a top or bottom, constructed from 22 mm thick planks that measure 278 x 256 x 160 mm on the outside and 234 x 225 x 160 mm on the inside. A 6 mm deep by 9 mm wide groove is carved along the top of the front and rear planks to serve as a resting place for the frames. The bottom board and the lower extremities of the frames are separated by approximately 6 mm. An entrance aperture measuring 88 mm by 9 mm is located on the lower side of the front plank.

Brood frame: Self-spacing (i) Height of side bar: 144 mm; width at top: 28 mm; width at bottom: 12 mm; (ii) Side bar breadth: 22 mm; length: 250 mm; thickness: 3 mm; (iii) Inner length of frame: 206 mm; Inner height of frame: 144 mm. When two frames are stored side by side, a 6 mm clearance is guaranteed and an extension of 3 mm is provided on either side of the side bar. A brood chamber contains seven frames.

Super and super frame: It is 78 mm tall, but it is the same length and width as the brood chamber. The interior height of the super frame is 62 mm, yet its external measurements are the same as those of the brood frame.

Top cover: On both sides are planks that slope. To allow for ventilation, a wire gauge-fitted 87 mm square aperture is constructed in the low ceiling plank. The required draught is provided via two holes in the top's front and back planks. It is important to provide a space of roughly 6 mm between the ceiling plank and the frames underneath. Light, well-seasoned, high-quality wood should be used to make beehives. The wood that is utilized shouldn't smell strongly.

Tectona grandis, Pinus excelsa, Kail, and *Toona ciliata* are a few of the woods that are appropriate for the job. To shield the wood from weathering agents, the outside of the hives should ideally be painted aluminum or white. To allow for interchangeability inside the apiary and within the specific region of the nation, the sections of the hive should be cut precisely.

Smoker: The smoker is a fire-starting metal cylinder that emits smoke. A bellow is linked to the smoker so that air may be blasted into the fire. To calm the bees and reduce their tendency to sting, controlled smoke is poured into the hive from the nozzle.

Honey Extractor: It is made up of a metal drum with a mechanism that rotates centrifugally to remove honey from the frames. With the aid of a handle, four frames full of honey from the super chamber can be placed inside the extractor at once to be extracted. The honey is then extracted through a pipe attached to the extractor's bottom and collected in jars, cylinders, etc. After extracting honey, the combs can be put back in the super chamber because using the extractor does not harm them. A two-frame extractor is also available, which is useful for home and small-scale beekeeping because it can only hold two frames at a time.

Comb Foundation Sheet: In the wild, bees create new combs by building parallel combs that are fastened to the hollow or box roof using beeswax that they secrete. The combs might be constructed obliquely, at right angles, or in the direction of the entrance. Straight combs must be incorporated into the frames of the moveable frame hive in order for the bees to maintain the proper distance from one hive to the next while being moved. Wax sheets are used in the comb foundation mill to prepare these comb foundation sheets. Since we are providing the basement, these comb foundation sheets will save the honeybees' energy as they create their combs.

Decapping Knife : Once honey is stored in honeycomb, bees seal the cells; therefore, in order to extract the honey using an extractor, we must use a decapping knife to open the sealed section. The knife used for decapping could be an electric heated knife or just a regular plain steel knife.

Hive Tool: It is a flattened iron tool with hammered edges that is used to scrape extra comb and bee glue from different areas of the hive as well as to pry apart the frames inside the hive.

Bee Veil: To protect the face against bee stings, it is worn over the face. It ought to be constructed using black light material, like nylon netting, to improve the image quality. It is important to ensure that veils fit tightly around the hat and onto the shoulder, with ample room remaining between the veil and the wearer's face.

Gloves: They provide a great deal of protection because handling honeybees usually results in stings on the palms and fingers. They help beginners gain confidence and can be made of rubberized cloth or strong canvas.

Bee Brush: Before extracting honey from honeycombs, bees are frequently removed with a bee brush, also known as a whisk broom.

Feeders: Beekeepers use a variety of feeders to provide sugar syrup to their hives. The division board feeder is a practical tool. It is a wooden trough with conventional frame dimensions and shoulders designed so it may hang in the hive like any other frame with a wooden strip acting as a float. Another excellent kind of feeder is a tin filled with sugar syrup and having holes in the lid.

Queen Cell Protector: A piece of wire twisted spirally forms the structure's cone shape. Around a queen cell, it fits. Before being accepted by the bees, a queen cell that needs to be transferred from a queen right to a colony without a queen is frequently shielded by a queen cell protector.

Division Board or Dummy: The wooden partition functions as a wall that bees may move and reduces the size of the brood chamber, allowing the hive to remain cool and safe from weather and other threats.

Embedder: This is a little instrument featuring a round or spike on top. It is employed to secure the comb foundation sheet to the frame's wires. In order to strengthen and reinforce the comb, electric wire is also utilized for this purpose.

Drone Trap: The drone trap is a rectangular box with an opening on one side. A queen excluder sheet is attached on the opposite side. There is room for worker bee movement at the bottom of the box. At the box's bottom wall are two hollow cones. Drones trying to enter the enclosure through the cones become stuck. The bees can pass out via the narrow end of the cone, but it's not big enough to draw their attention or encourage them to re-enter. To lessen the number of drones within the hive, this device is placed at the entrance.

Pollen Trap: Pollen-trapping screen scrapes pellets from the legs of foragers that are returning. It lies at the entrance to the hive. The collected particles of pollen land in a receiving tray that resembles a drawer.

CHAPTER 8

Bee Products

Honeybees offer humanity with highly profitable products in addition to their vital pollination services. The main gift that honey bees offer is honey, as the name implies; they additionally provide bee venom, pollen, royal jelly, propolis, and beeswax.

Honey

When plant nectar and sugar deposits are collected, processed, and stored by honeybees in their honeycomb, they then alter their contents to make honey. Both floral and extrafloral nectar types affect the quality of honey. Honey can be classified into two categories: uni- and multifloral. Honeys can be categorized into groups according on the type of flower they originate from, such as Brassica, eucalyptus, litchi, and berseem, despite being rare. Carbohydrates, usually in the form of glucose and fructose, are abundant in honey Water, which also contains a wide range of other sugars, acids, vitamins, proteins, and minerals, is the other main ingredient of honey. Honeybees seal the honey inside comb cells once they have reduced the excess moisture to less than 20 percent. There are two ways to gather honey: to get rid of impurities like pollen and larvae, you can compress the combs of *Apis dorsata*. But in order to filter out impurities, honey from domesticated bees (*A. mellifera and A. cerana*) is gathered using centrifugal force and a honey extractor.

Depending on whether it crystallizes or remains liquid, honey can be sold to consumers as either liquid or granulated. The processing of honey involves two phases. To kill the yeast cells that cause honey to ferment, the honey is first indirectly heated in a water bath and kept at 60 C for 30 minutes. The honey is then cooled, filtered through a two-layer cloth filter while still warm, and bottled in glass bottles when its viscosity has decreased.

Composition of Honey

The compounds that makeup of honey is extremely intricate. Although sugars make up the majority of the substance, honey also contains trace amounts of acids, minerals, vitamins, enzymes, and antibiotics. The floral source affects the solids in honeys. When the honey reaches about 20% moisture content, honeybees seal it in comb cells after putting it through the ripening process. Compared to honey from the Western and Eastern Ghats, the Northeast, and regions with extremely humid weather, honey from the scorching plains of north India has less moisture content. In these regions, honeys have a thinner consistency and can include up to 23% moisture.

Due to the hygroscopic nature of honey and its ability to collect moisture, the moisture contents may even alter after extraction during storage. Although there are many different kinds of sugars in honey, levulose (fructose) and dextrose are the main reducing sugars. About 70% of the total solids are made up of the two sugars. Honey's sucrose content ranges from 0% to 5%. Maltose, a reducing disaccharide, is also present in greater concentrations, ranging from 3–16%. Although this disaccharide may not be present at first, it does accumulate during storage, most likely as a result of enzymes and acids converting other sugars. Numerous additional less common sugars, including isomaltose, trehalose, genitiobiose, maltulose, panose, turanose, and others, have also been detected using precise analytical techniques.

Only 1-2% are made up of these complex carbohydrates. Honey has an extremely diverse mineral complex that includes up to 1% of solids in the form of potassium, calcium, phosphorus, sodium, magnesium, manganese, copper, sulfur, silica, silicon, and iron. The vitamins thiamine (Bl), riboflavin (B2), nicotinic acid, vitamin K, folic acid, biotin, pyridoxine, pantothenic acid, ascorbic acid, and carotene are among those found in honey. Vitamins are lost while being stored. The following lists the ascorbic acid, thiamine, riboflavin, and niacin losses in several unifloral honeys.

Physical Properties of Honey

Honey is standardized not only by its chemical content but also by its physical characteristics. These characteristics also testify to the honey's purity. Honey absorbs moisture from the surrounding air due to its hygroscopic nature. In high humidity environments, honey will tend to absorb moisture from the air if improperly stored. As the moisture content rises, this happens in the honey's surface layer. Honey's hygroscopicity is dependent on its moisture and sugar content. The moisture content of honey is calibrated to a particular relative humidity level. For example, 66% relative humidity balances the 21-23% moisture content of honey. If honey is exposed to air with a lower relative humidity, it may lose moisture. On the other hand, honey can be kept in conditions that are exceptionally humid without becoming wet and potentially fermenting.

Hygroscopicity has practical uses for beekeepers. Honey's viscosity determines its flowability because it is a viscous liquid. The range of the total reducing sugar was 63.62 to 68.1%, and the range of the moisture content was 16.22 to 18.87%. The mineral content of Apis dorsata honey was greater (0.33) when compared to samples from other honey bee species (0.20-0.24%). Fiehe's test produced a negative result for each sample. In India, honey has a moisture content ranging from 19 to 25%. For glucose, the range of values was 35.12 - 38.04%, while for fructose, it was 37.27 - 40.51. Viscosity measurements can be used to assess moisture content. Honey must be heat-treated in order to be handled since heat reduces viscosity.

But viscosity is also influenced by protein concentration, which is influenced by the source of nectar. Honeys with more protein usually have a higher viscosity.

Honey's specific gravity can be calculated by dividing its weight by the weight of a volume of water that is comparable to it. The specific gravity of pure honey should be in the range of 1.35 to 1.44. While there are many ways to determine specific gravity, temperature is typically given. In contrast, the ratio of light velocity in honey to air velocity is known as the refractive index of honey. Honey can have its moisture content measured with a refractometer.

Measuring these two characteristics will yield quick estimations of the moisture content. Honey smells and tastes unique. This aroma is contributed by floral nectar, which makes it an essential physical attribute for assessing honey. The wrong storage containers might give some metallic flavors to honey since it can collect odors from the outside world. Pressed honey and apiary honey must meet purity standards set by the Indian Standard Institution.

Fermentation of Honey

Increased level of moisture Honey can undergo fermentation. This is especially likely to happen if the honey crystallizes in an uneven pattern. At the bottom, the honey starts to granulate as the dextrose crystallizes to form solids. The top layer of honey gets diluted when the amount of sugar in it declines. A range of sugar-tolerant osmophilic yeast species are responsible for the fermentation of honey.

The optimal temperature range for honey to ferment is between 11 and 21 degrees Celsius, and as temperature rises or falls, honey's inclination to ferment reduces. Honey fermentation results in the production of carbon dioxide and alcohol. If the alcohol is converted to acetic acid and water further, all of the honey in the container will ferment. Because of its high acidity, which can vary from 1.5 to 3.1 milli equivalents, fermented honey has a sour flavor. Fermentation of honey is halted when it is heated to 64 C for around 30 minutes, destroying its yeasts. Fermentation is unlikely to happen if the honey granulates quickly and uniformly.

Honey Crystallization/ Granulation

The most serious challenge being faced by Indian beekeepers is granulation of honey. Customers do not accept granulated honey because they think it is a sign of contaminants or adulteration. Granulated honey is also difficult to handle. Some honeys crystallize even in combs; *Brassica* honey demonstrated this by having some of the combs left partially open by the bees. Some honeys soon start to granulate after being extracted. This makes it impossible to extract honey, and these combs are only good for feeding colonies when there is a famine. Most customers aren't aware that honey is a super-saturated sugar solution.

The two main sugars in honey are fructose and dextrose. Dextrose crystallizes easily, which is why honey has a built-in tendency to granulate. Honey's granulation is also influenced by its moisture content since fructose dissolves in water more easily than glucose. The fructose to glucose ratio shows the rate at which honey granulates. A higher L/D ratio honey will granulate more slowly. *Carvia callosa* honey has an L/D ratio of 1.43, is composed of 38.29% levulose and 28.17% dextrose, and it never granulates. *Strobilanthes ixiocephalus* honey granulates quickly and has a ratio of 1.038. *Syzygium cumini* (jamun) has a honey ratio of 1.10, however as the honey does not crystallize, this ratio is not a reliable indicator of the granulating characteristic.

Thus, it appears that crystallization is influenced by a number of parameters in addition to the L/D ratio. Dextrose (D) to moisture ratio (W) has recently been proposed as a more precise measure of granulation. Higher D/W ratio honey is predicted to granulate more quickly, yet this is not always the case—jamun honey is another example. However, two criterion is sufficient for generality. Any foreign particles, such as dust particles, pollen grains, wax fragments, and even air bubbles, function as granulation nuclei. These nuclei are where the granulation begins. Honey can granulate quickly and uniformly. In these circumstances, the honey has little possibility of fermenting because the granules are fine. However, some honeys granulate slowly, causing the crystals to fuse together into larger lumps that granulate coarsely. While the top layer of honey remains liquid, the lowest layers of the container granulate. Fermentation takes place when the liquid layer on top gets diluted. Granulation is delayed when honey is heated because it kills the yeasts and dissolves all of the dextrose granules.

Low temperatures, between 0 and 5 C, also reduces crystallization. Approximately 14 C is the ideal temperature for crystallization. This characteristic provides beekeepers with a crucial hint about the ideal honey storage temperature. Honey's antioxidant qualities Antioxidants are compounds that shield the body from the harm that free radicals can inflict. Free radicals are highly reactive substances that interact with nearby molecules to modify them.

Since honey is a good natural antioxidant, it scavenges free radicals by either preventing them from forming or neutralizing them. Honey is a nutritious food that is good for human health. The biochemical properties of honey are significantly influenced by its content and source. Using spectrophotometric tests—the ferric reducing antioxidants assay (FRAP assay) for total antioxidant activity, the 1, l-diphenyl-2-picrylhydrazyl (DPPH) assay for antiradical activity, and the florimetric method, specifically the oxygen reactive antioxidant capacity (ORAC) for the antilipoperoxidant activity—the study focused on the antioxidant/radical scavenging capacity of various honeys of different flora and geographical origin.

Honey's phenolic and other chemicals are what give it its antioxidant and free radical-scavenging properties, which are good for human health. Honey processing: The honey found in comb cells is free of all external materials save pollen grains and has a pleasant flavor and scent. The beekeeper himself may cause the honey's quality to decline during extraction and handling for marketing. Honey incorporates comb wax fragments during the extraction process. A portion of the scent is lost during the thin-film extraction of the honey from the comb cells with the aid of extractors. Honey may also get some airborne dust added to it during the extraction process. The quality of honey will further decrease under unfavorable storage conditions, such as improper containers, high storage temperatures, etc.

Honey Processing

The only approach used to process honey is heating. However, heat treatment might have negative effects if it is applied carelessly. There are several ways that heat affects honey, which is covered below:

1. Honey sugars break down to create hydroxymethyl furfural. Although heating is the usual cause, this transformation also occurs with extended storage at comparable higher temperatures.
2. Heat destroys the enzymes in honey.
3. Heating makes anything easier to handle by lowering its viscosity. The ideal temperature range for moving honey in order to accomplish this goal is 38 to 43 C.
4. The process is helpful in eliminating yeasts from honey. If it isn't polluted again, this honey won't ferment. By heating the honey to 71.1 C for 0.4 minutes, 65.5 C for 2.8 minutes, or 60.0 C for 22 minutes, yeast can be removed from it. When pasteurizing honey, it takes some time for the honey to reach the proper temperature.
5. Dextrose crystal nuclei can be found in honey. Even though these minuscule nuclei may not be seen at first, they are what cause honey to crystallize.

Granulation cannot be prevented; however it is delayed because the heating destroys the crystal nuclei. The honey will flow easily if it is first warmed to between 38 and 43 C. After that, the heated honey is transferred to a holding tank at around the same temperature. Wax and other foreign particles are eliminated from the holding tank by rising to the surface. There is a paddle-style agitator included with the holding tank. When immersed in honey, the paddles rotate at a speed of 10 revolutions per minute. The stirring aids in the blending and homogenization of honey. To prevent air bubbles from getting into the honey, the agitator is turned slowly. The strainers come next; they are offered in a series with progressively smaller mesh sizes, with a minimum of about 80 mesh/inch. All particles are removed by these strainers. The high pasteurization temperature (66 –70 C) will melt any wax particles that are not removed by straining, giving the honey a hazy look.

After being strained, the honey is run through a tank to kill the yeast by heating it to 63 –65 C for five to seven minutes. This is accomplished through flash heating, which involves heating honey in thin layers and is necessary due to its low conductivity. In order to prevent quality degradation, the highly heated honey is promptly cooled. For small-scale beekeepers, filtering and warming honey is the only practical way because they cannot afford more complex processing setups.

Storage

It is recommended to use glass or stainless steel storage containers for liquid or crystallized honey, or to cover them with food-grade plastic, paint, or beeswax. Nothing that gives the honey an odor should be permitted. When recycling specially infused containers, it's important to make sure they are spotless and don't contain any lingering smells. All types of odors are easily absorbed by honey; for example, odors can easily travel through a beeswax layer and into the honey. Bee products, even after being coated with paint, plastic, or beeswax, should never be stored in containers that were once used for hazardous chemicals, oils, or petroleum products.

Large enough openings in wholesale containers are necessary to make it easy to remove crystallized honey. Lids must be airtight, and items should be stored away from heat sources and light (ideally) to prevent moisture from getting inside. Additionally, the majority of honey-containing products should be packaged specifically to prevent excessive moisture damage. For instance, single servings of liquid or ground product should be placed in paper and plastic envelopes coated with aluminum foil; baked items should be packaged in moisture-resistant clear plastic bags; and caramels should be wrapped in waxed paper or plastic. For storage rooms, the ideal temperature is about 20 C and the ideal relative humidity is less than 65%. Honey that is kept in storage over 25 C eventually loses quality due to chemical and enzymatic changes.

Honey is considered a stable product since it is not damaged by the germs and fungus that usually cause food to decay. However, honey-containing products are highly attractive to these types of organisms, therefore careful storage and post-production protection against recontamination are essential, as is pasteurization (heat-stabilization) or chemical preservatives based on product specifications. With appropriate packaging, storage, and marketing techniques, together with timely marketing and consumption, the usage of preservatives will be reduced or completely eliminated. Untreated honey is largely susceptible to fermentation regardless of its form—liquid or crystalline. Because of this, in order to avoid fermentation, storage conditions must either prohibit additional moisture absorption or use low temperatures.

Even while honeys are not susceptible to yeast deterioration, they might nonetheless experience additional progressive alterations due to chemical and enzymatic action. These changes include the loss of organoleptic qualities including color, taste, and aroma as well as physiologically active compounds (inactivation of enzymatic and antibacterial activity). There may also be notable changes in the sugar composition, with a rise in disaccharides and other complex sugars and a decrease in simple sugars.

Increased acidity and a greater concentration of hydroxymethyl furfuraldehyde (HMF) are two more modifications to honey's natural composition. All honeys go through these changes, however they do so at different rates based on the starting composition (greater moisture content and lower pH honeys experience changes more quickly) and storage temperature (higher temperatures cause changes to occur more quickly in honeys). The alterations happen even faster when various processing methods are heat-treated. Damaged honey loses some of its nutritional and organoleptic properties even if it doesn't become hazardous for humans.

In light of this, nearly every nation has laws limiting how much honey can "age" (or degrade) before it can be used for food. Heat and sunlight, especially in the UV range, can degrade honey either over a long period of low exposure or in brief bursts. Although some degradation is unavoidable, it should be kept to a minimum. UV radiation destroys glycooxidase and reduces antibacterial action significantly. The diastase half-life in honey at various storage temps. There are several factors that could affect how long honey lasts, thus it's difficult to pinpoint a precise limit. As a result, diastase and HMF are used as indicators of potentially dangerous processing or storage methods for honey. As a result, honey may suffer more damage if the half-life decreases or if the rate of diastase clearance increases.

Beeswax

Wax is one of the precious substances that bees synthesize. The worker bee's wax glands start to produce wax when it is 14–18 days old. Normally white, the wax may take on a different color when used by bees to build their nests, depending on the source of the pollen. The specific gravity of beeswax is 0.95, and its melting point is 65 C. Beeswax has a complex mixture of monoatomic alcohol esters, 70.4–74.7 percent fatty acids, 13.5–15.0 percent, and 12.5–15.5 percent saturated fats. The typical content of honey Higher sugar content: 17.2, pH 3.91, nitrogen 0.041, fructose 38.2, glucose 31.3, sucrose 1.3, maltose 7.3, 0.43 free acids, 1.5 acid total, 0.57, lactone 0.14, ash 0.169, and 4. Diastase value: 20.8 M for 51 hydrocarbons.

The usual source of wax production is broken combs. Wax can be extracted in two ways: the most common method in India is to use a hot water bath, but you can also use a solar extractor, which melts wax with the help of solar radiation. Wax is used to make comb foundation sheets, candles, polishes, furniture, the pharmaceutical industry, and the perfume industry because it sticks to skin more readily. It is also a necessary ingredient in lipsticks, rouges, and cold creams, among other cosmetics.

Bee Venom

The vital substances found in bee venom are proteins and peptides. Numerous documents suggest that apitherapy using bee venom was practiced in China, India, Egypt, Babylon, and Greece. Bee venom is contained in the venom sac and is administered by stinging. Bee venom can be extracted using a venom extractor that uses a low electric current. The venom is gathered on a glass plate at the bottom when the irritated bees try to sting due to the current. In the last thirty years, the USA, the world's largest venom generator, has produced only 3 kg of venom. In India and certain other Asian nations, apitherapy is a traditional method of treating joint pain that involves holding a bee between the thumb and index finger and allowing it to sting. The venom prepared by the procedure outlined above can be utilized in subcutaneous injections. On the affected areas, use an ointment comprising apitoxin, Vaseline, and salicylic acid (1:10:1). Salicylic acid softens the skin and encourages penetration.

Propolis

Propolis is a dark-colored, sticky substance that is used to seal hive cracks and fissures and assist honey bees adapt and build nests. Propolis can range in color from light yellow or brown to dark brown in temperate climates, often with a reddish undertone. Propolis often becomes darker the longer it is stored in the hive. The color of propolis varies greatly depending on the plants and trees that are utilized to collect it. It can be used to treat wounds, infections, dermatitis, and cancer. It is an effective fungicide and disinfectant. It prevents bacteria, fungi, and yeast from growing. Among the Apis species, only *A. mellifera* is known to forage for propolis. The tropical stingless bees accumulate a resinous substance that resembles propolis, which they use to seal their hive and create storage receptacles for honey and pollen.

Royal Jelly

Royal jelly, a fluid produced by honey bees, serves to feed larvae and adult queens alike. It is secreted by glands in the nursing bee's hypopharynx and fed to all larvae in the colony, irrespective of sex or caste. In the process of producing new queens, the laborers build unique queen cells. The larvae are fed enormous amounts of royal jelly through these cells. This type of feeding initiates the development of the queen's morphology, including the fully developed ovaries necessary for egg production.

Royal jelly, secreted by glands in the heads of worker bees, is supplied to all bee larvae, regardless of whether they will become workers (sterile females), queens (fertile females), or drones (males). After three days, royal jelly is no longer fed to worker and drone larvae; however, queen larvae continue to receive it throughout their development. Royal jelly is composed of 67% water, 12% protein, 6% fatty acids, 1% simple sugars (monosaccharides), and 3.5 percent 10-hydroxy-2-decenoic acid (10-HDA). It contains trace minerals, antibacterial and antibiotic components, pantothenic acid (vitamin B5), pyridoxine (vitamin B6), and trace amounts of vitamin C, but not the fat-soluble vitamins A, D, E, or K. The royal jelly has a creamy white texture.

It is highly acidic, rich in protein, carbs, vitamins, RNA, DNA, and fatty acids, and generated by nurse bees between the ages of 6 and 12 days. It also increases energy and vitality, which makes it an especially healthy snack for people. Royal jelly is a rich source of amino acids, including arginine, aspartic acid, glutamic acid, glycine, isoleucine, lysine, methionine, phenylalanine, tryptophan, tyrosine, and serine. Eight of the essential amino acids required by humans are found in royal jelly. It also contains the vitamins A, B, and C, iron, copper, phosphorus, silicon, and sulfur. Royal jelly is generated via a procedure called grafting, in which fake queen cell cups made of pure wax are attached to a brood frame formed of bars containing tiny blocks of wax. After that, the larvae will be kept in the queen cell cups inside the hive until they are one or two days old.

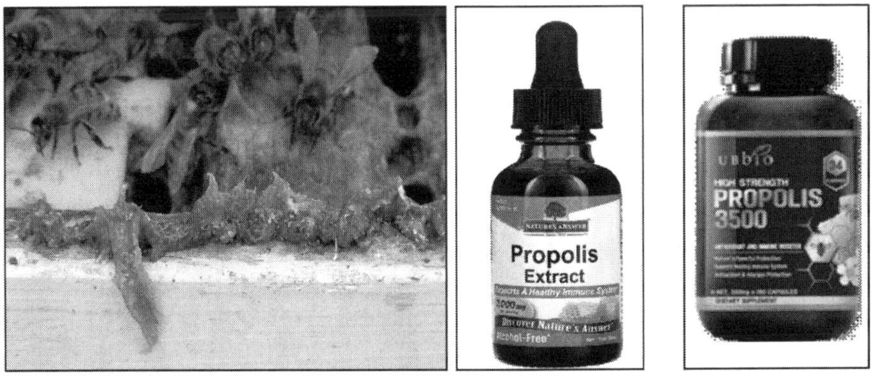

Propolis

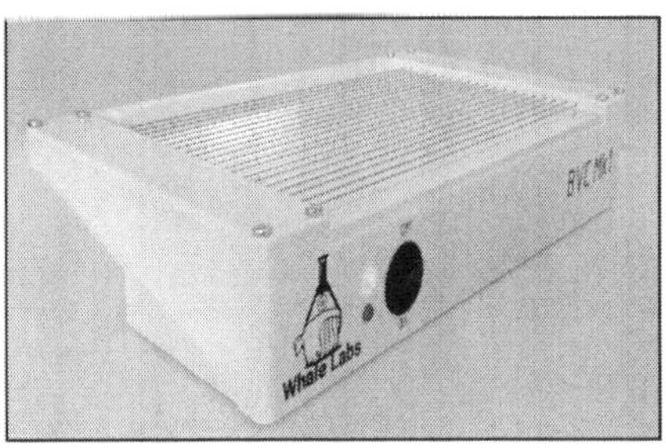

Bee Venom Collector

Bee Venom

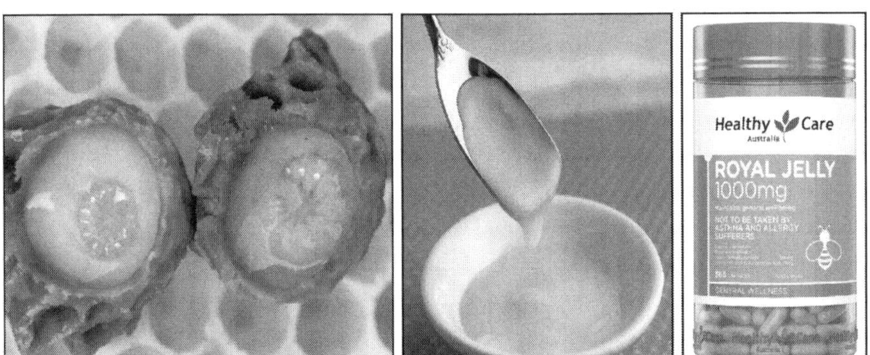

Royal Jelly

The nurse bees feed the larvae royal jelly, which is obtainable. Honey bee workers and queens exhibit one of the most amazing examples of phenotypic variation under environmental control. Even if two larvae with the identical DNA were raised to become workers and the other to become queens, the two adults would differ substantially in terms of morphology and physiology, longevity, and ability to reproduce. Female workers have rudimentary, inactive ovaries and are functionally infertile, while queens, the largest members of the female sexual caste, have gigantic, working ovaries. The protein royalactin is specifically responsible for controlling the developmental difference between queens and workers epigenetically by differential feeding with royal jelly. A female larva that is intended to become a queen is given large quantities of royal jelly, which initiates a sequence of molecular events that ultimately result in the development of a queen.

Scientific Beekeeping & Value Addition of Honey and Bees Wax Products by Elp Bee Keeping Students of Dr Ysrhu - College of Horticulture, Venkatramannaudem.

Beekeeping Unit was established at College of Horticulture, Dr YSR Horticultural University, Venkataramannagudem, West Godavari District, Andhra Pradesh in 2017. The main objectives of the beekeeping unit are as follows:

1. Establishing Nucleus stock development unit.
2. Multiplication of nucleus stock into bee colonies.
3. Development of Information centre for scientific bee keeping.
4. Collection and storage of honey and other bee products and quality control
5. Development of bee by products.
6. Revenue generation through sales of honey and bee products.

Major Achievements

Activity/Awards	Achievements
VALUE ADDITION OF HONEY AND BEES WAX AND PROFIT GENERATION	ELP beekeeping students are trained on beekeeping and preparation of value addition products of honey and bees wax viz, Honey, Lip balms, Bees wax candles, Mosquito repellers, Shoe polish, Cockroach baits, pain balms, honey cakes , Room freshener, crayons, lipsticks, Dry fruit honey laddoo, Eye liner, honey lollipop, Honey chocolates, Honey peanut butter sweets, Honey Ragi pudding, Honey, fruit custard etc.
AAVISHKAR AWARDS - 2022	ELP Beekeeping students. G. Param jyothi and P.Sharon received 3rd position in AVISHKAAR-2022, AP STATELEVEL INNOVATIVE Competitions organized by APCOST & AICTE ON 04 & 05TH March 2022 for preparing BEES WAX CRAYONS held at AdiKavi Nannayya University, Rajamahendravaram

EXCELLENCE IN INNOVATION AWARD – 2020	Dr N. Emmanuel, Associate Professor (Ent) & ELP Beekeeping module in-charge has Received EXCELLENCE IN INNOVATION AWARD – 2020 from DR B.V.VASANTH RAJ DAVID FOUNDATION , on 5th December 2020 at Le palace hotel, Chennai, for his Contribution to an Innovative Experiential Learning Programme –preparing multiple indigenous honey and beeswax value added products

Scientific Beekeeping & Value Addition of Honey and Bees Wax Products by Elp Bee Keeping Students of College of Horticulture, Vrgudem, Dr. Ysrhu

COSMETICS	LIP BALMS
	MOSQUITO REPELLERS
	PAIN BALMS
	HAIR CREAMS
	ALOE HONEY SANITIZES
	EYE LINER
	FACE FRESHNER
	HONEY BEES WAX GOAT MILK SOAP
	DEODARANTS
HOME NEEDS	BEES WAX CANDLE
	SHOE POLISH
	WOOD POLISH
	COCKROACH BAITS
	ROOM FRESHENER
	CRAYONS
	INSECT KEY CHAINS
HONEY FOODS	IMMUNITY BOOSTER DRINKS
	HONEY NUTTY CHOCOLATES
	HONEY CAKES
	PEANUT BUTTER SWEETS
	DRY FRUIT HONEY LADDOO
	HONEY RAGI PUDDING
	HONEY FRUIT CUSTARD

List of Honey and Bees Wax Value Added Products Prepared by Elp Bee Keeping Students of College of Horticulture, Vrgudem, Dr Ysrhu

G.Param Jyothi of COH, VRGudem received 3rd position in AVISHKAAR- 2022, AP STATELEVEL INNOVATIVE Competitions organized by APCOST & AICTE ON 04 & 05TH March 2022 for preparing BEES WAX CRAYONS held at AdiKavi Nannayya University, Rajamahendravaram

Dr T.Janakiram , Hon'ble Vice Chancellor of Dr YSR Horticultural University congratulating the AVISHKAAR- 2022 award winners

Dr N.Emmanuel, Associate Professor (Ent) & ELP Beekeeping module in-charge Received EXCELLENCE IN INNOVATION AWARD – 2020 from DR B.V.VASANTH RAJ DAVID FOUNDATION, on 5th December 2020 at Le palace hotel, Chennai, for his Contribution in preparing multiple indigenous honey and beeswax value added products through ELP Programme

CHAPTER 9

Seasonal Management and Colony Maintenance of Bees

The vegetation and weather have an impact on management techniques. The beekeeper must adjust to the conditions that are present in his area at various points in time. The primary things to keep in mind are the presence of enough nectar-producing plants within the bees' reach and ideal weather that allows the bees to fly outside and gather food for an adequate amount of time each day.

Management During Winter

1. The winter months are frequently a busy time for bees in the plains when conditions are mild. For example, the mustard and related crops bloom in the plains during the winter months of November to January, giving the bees access to an abundance of nectar and allowing them to gather enough extra honey.
2. The bees should be provided with enough protection throughout the winter months on hills where conditions are harsh. They must be winterized and packaged well before the first hard frost.
3. An effective method for accomplishing this is to cover the entire hive—the top, bottom, and all six sides—with dry grass and secure it with threads. Permit the bees to fly out by keeping a gap open in front of the entrance.
4. On days when it rains a lot and snows, they ought to be covered in leaves.
5. The hives need to be adequately shielded from the chilly winter breezes. There should be a windbreak.

Management During Spring

1. In the lowlands as much as in the hills, spring is a crucial time.
2. When bees are packed, they should not be unpacked until the weather gets warmer.
3. Once the packaging has been opened, inspect the colonies and feed them either starving food or simulative food based on each colony's requirements.
4. The bees raise a lot of brood in the spring, therefore they require a lot of water. Watering holes should be situated somewhat close to the beehives.

Management During Summer

Summer provides no problem on the hills. But in the plains, summer is the difficult part of the year for the bees.

The heat and lack of humidity have adverse effects on the bees and their activities.

During the hot part of the year, the following steps need to be taken:

1. Keep the hive in deep shade
2. Feed a watery sugar solution
3. Sprinkle water over and near the hive every morning and evening

Management During the Monsoon

1. This is generally the difficult period on the hills.
2. Bees remain inactive because of constant rain.
3. Brood rearing dwindles and the colony becomes weak.
4. Wax-moth activity is at its peak during this period.
5. Wasps are serious on hills management during pollination
6. Individual crops vary in their bold requirement of bee colonies for pollination. The number of bees required depends on the number of flowers available in a given area for pollination. Generally 3-9 bee colonies of good strength are needed for one hectare of crop.
7. The basic need for the management of colonies for pollination is to keep the colonies in peak foraging activity. This is possible when there is peak brood rearing
8. Farmer himself can learn bee-keeping and try to have a few of his own bee colonies rather than depend upon bee-keepers
9. By this way a farmer can become self-reliant

Colony Maintenance of Bees

A colony of honeybees is really a family in which the association between the 3 castes for common benefit is so highly organized and their activities so efficiently coordinated that the individuality of the single bee becomes merged with the complex social organization. The qualities, which help honeybees to lead a successful social life, are given below.

The colony has a single queen. Due to her long span of life she develops a parental case and love for the other members of the colony, thus strengthening the foundations of family life.

1. In a Colony, each individual has to work for the mutual benefit of others. There is a remarkable degree of division of labour among the nurse bees as well as among the foraging bees.

2. The bees in a colony have 'hive mind' and live together by the 'hive odour' peculiar to each hive.
3. The habit of storing food in the comb for long periods helps the family life.
4. Co-ordination of activities.

Colony Division

Honey bee colonies which are not very strong and be spared from honey production can be divided into 2 or 3 divides each divide is given a new queen. These divides grow in spring and summer and colonies produce surplus honey only during next season. One or two combs with bees and brood can be removed from strong colonies without impairing their production. The bees from two colonies can be united and made into a new colony by giving a queen. In hills, the winter is severe and there is lack of flora, the bees remain confined to their hive for most of the time. The bees have also been facing the problem of maintaining nest temperature. At the onset of spring, the colonies emerge in considerably weak condition. It should be the earliest attempt to examine the colonies on a bright, warm and calm day to assess the condition of the colony, working of the queen, amount of brood present, honey and pollen stores and to clean the bottom board debris accumulated during winter. It is a useful management to give a stimulant feeding to colonies when very few spring flowers have blossomed. The stimulant feeding with sugar and protein supplement (30 to 40%) will help the colonies to rear more brood and raise greater foraging force to avail spring and summer flow. Thus feeding also raises the morale of the bees. It is an established fact that colonies which receive stimulant feeding produce more honey. The examination of colony should be done carefully and quickly because robbing is easily induced. All manipulations should be stopped once the robbing becomes apparent and reduce all entrances so that colonies are able to guard against the invasion of robbers.

During spring, bee colonies go 'all out' to rear brood and invest all resources in increasing their strength. Queen lays more vigorously after winter egg laying rest. More drone combs are added for expanding brood nest. If the queen is working unsatisfactorily, that she is laying sparingly and/or laying drone eggs, effort must be made to replace the queen at the earliest opportunity. It may be pointed out that very weak colonies desert their hives if disturbed unnecessarily and are easily robbed out if they are fed syrup without great care. It is a wise practice to unite them with others and follow the golden rule, "always help first those colonies that need the least help, leaving the weakest to be helped the last". Colonies which are just below average in strength may be helped by giving them each a frame or two of capped brood from strong colonies. Another good practice is to equalize the strength of the colonies to avoid disparity so that all the colonies require the beekeepers attention at about the same time and respond equally to the treatment. During early spring, the weather is uncertain and beekeepers are warned against over expanding the brood nest and dividing it into two or more part by insertion of empty comb or comb foundations because there can be chance of the outlying brood being left unattended by the worker bees. Consequently, the neglected brood gets' chilled' because of treacherous weather that can prevail during early spring. But on the other hand, shortage of frames of stores at this time of the year cripples the colony for the whole season and the colony is not fully prepared to avail good honey flow.

Similar management practices in spring are needed under mid hill and sub mountainous conditions. But in plains of north India, the conditions are different in spring. In most areas extensive Brassica flora is available from October to January. The bees get 3 to 5 hr for foraging during a day and instead of depletion of strength, there is some build up. Good colonies attain yielding strength and honey extraction is possible in January. The precautions against extension of brood nest, stimulant feeding etc. can be relaxed. Similarly, bees in southern peninsular India do not face weather problem. Spring is also a swarming season. Swarm prevention and control measures should be taken. Sharma and Kumar (2001) while evaluating the appropriate time for *Apis mellifera* colony division at Hisar during spring and autumn reported that February/March and November/December are suitable for colony division in Hisar, Haryana. Honey flow period Major honey flow period starts after winter and onset of spring season. Management of the bees is essential during these days. It is true that any amount of negligence during the honey flow will result in the poor management of the colonies ultimately leading to the crucial end of the colonies. To manage the colonies efficiently, the beekeeper should have the fullest knowledge of honey flow trends and the state of the colonies.

The principal function of the beekeeper during honey flow period is to keep the colony strength properly. In other words, he should ensure that the honey gathering instinct is dominant and that the instinct is not checked. Congestion in the hive must be avoided and surplus house bees are drawn to supers. A colony is particularly liable to have this trouble during honey flow because of rapid nectar income, and the beekeeper should do his best to remain ahead of bees providing space for honey storage. It must be remembered that great deal of space is required for the evaporation of honey than for its final storage. Therefore, drone combs should be provided liberally.

A second super should be added between the first super and brood chamber and not above the first super. Many times, the queen goes to the super chamber and lays eggs and honey extraction becomes difficult. At last, weeks before honey extraction, a queen excluder should be placed in between brood and super chamber, and queen is confined to brood chamber. The supers can be of full depth or half supers. Handling and various operations are easy with full depth supers because frames can be exchanged between the chambers. But the judgment to use the kind of super depends on the colony strength. *Apis mellifera* colonies attain sufficient strength before honey flow and full depth supers on Langstroth hives are used in India. In case of ISI hives for *A. cerana* full or half supers are used. In Jammu and Kashmir Langstroth hive is used even for Indian honey bee. In most southern beekeeping areas, the Newton hive is still in use, where halfsuper is common.

Colony should be examined once a week and frames full of honey should be removed to the sides of the super and such frames can be raised from brood to super chamber. The frames which are three-fourth filled with honey or pollen and one -fourth with sealed brood should also be taken out of brood chamber and in its place empty combs or frames with foundation is added. Bees draw comb foundation better and quickly in brood chamber. The frame with comb foundation should be placed next to the brood nest. Often the beekeeper is short of drawn combs and getting more combs raised would lower honey yield. In such cases, the combs which are completely sealed or two thirds capped may be taken out for honey extraction and returned to the supers after honey extraction. Such extraction of uncapped honey should be avoided because this unripe honey has higher moisture percentage and is liable to ferment. Honey extraction, after the flow is over, can induce severe condition not within the species but also between the species too. It is a common experience that *A. dorsata* robbing domesticated hive bees become beyond control and loss of colonies is the result. On warm days, bees are noticed to gather in clusters at the entrance. This is a sign of congestion and poor ventilation. This affects the honey gathering instinct. The situation should be remedied promptly by improving ventilation and removing the entrance rod placing or queen gate.

CHAPTER 10

Queen Bee Rearing

Honey bee colonies go through a number of circumstances in its natural condition that lead to the emergence of a new queen. Beekeepers may raise healthy new queens for their hives using an array of queen raising methods if they are aware of these triggers and how they impact honey bee behavior. One of the most important elements of a productive honey bee colony is a robust and energetic queen. She is a reliable source of new workers. She assists in directing the actions necessary to maintain the colony's health and productivity through her pheromones. But there are a few situations that will force a colony to make a new queen on its own.

Why new queens are raised

1. An aging or ill queen will habitually produce smaller amount and/or lower quality eggs. The egg laying pattern often turns into spotty. These and supplementary precise cues will cause the workers to create supersedure cells.
2. A supersedure cell is where the colony will raise new queens to replace the weakening queen. Hives that become too packed will often cause the queen to leave in seek out of a new abode. The queen generally takes more than half of the hive's workers with her. Whilst a colony senses their queen is preparing to swarm, they will turn out swarm cells to heave a new queen who will take her place after she is departed.
3. When a queen relinquishes the hive due to illness or injury, a colony will swiftly recognize her absence. They are able to perceive this as the levels of her pheromones rapidly drop in the hive. This will trigger workers bees to select numerous juvenile larvae and nourish them exclusively on royal jelly to create new queens.
4. Beekeepers become conscious of dipping levels of new brood in their hives may choose to requeen them. This makes certain that the colony has adequate workers foraging for nectar and pollen during peak nectar flow to maintain the hive well stocked for winter.
5. To shun swarming, beekeepers may prefer to split a hefty colony into two or more new hives. Supplying one or both of the new colonies with a new queen ensures a stable supply of fresh brood.

6. Launching a new queen derived from the larva of a highly productive colony can facilitate to improve the output of a less productive colony. Traits such as temperament, resistance to diseases and pests, seasonal population adjustments and levels of honey production are all affected by the colony's genes. Within a few weeks, workers from the new queen will replace the older population and the colony will exhibit the desired traits.

Analogous to other perspective of beekeeping, queen rearing techniques could vary significantly amongst beekeepers. Several of the most widely used techniques in use today can be divided into two broad groups. An actively laying queen is taken out of the main hive using the first approach. She is given artificial queen cups to lay her eggs in. In the second kind, a process called grafting is used by the beekeeper to remove freshly hatched larvae from the brood comb to prepared queen cups.

The Queen

The beekeeper might influence the productivity, temperament, and behavior of the colony through the queen and her offspring. Queen bee raising techniques have been developed to enable beekeepers to start new colonies or replace old or unwanted queens in their colonies with healthy stock. A queen bee's huge pea-sized thorax and long, tapering abdomen make her slightly larger than a worker bee. Her reproductive organs are fully functional, unlike the worker bees', which makes her different from the other female bees.

The colony typically consists of one mature queen at a time, and she is in charge of reproduction. A fully grown queen can lay up to 2000 eggs in a single day in phases, although under ideal circumstances, she typically lays 700–1000 eggs daily. Although a queen can survive for several years, in commercial settings, her vigor—which is correlated with the health of the colony—means that she is normally replaced every one to two seasons. Six to ten days after exiting from her cell, she mates with up to twenty drones on the nuptial flight. Once she starts laying eggs, she won't mate again.

Natural Queen Replacement

A colony of honey bees in the wild, or in an undisturbed domestic hive, will raise queen bees for three reasons: to substitute a lost queen, to reinstate the queen when swarming, and to replace or supersede a failing queen. When raising queens, beekeepers will manipulate a colony to duplicate one of these three natural circumstances.

Bees prepare emergency cells in the event that the regular laying queen is unintentionally killed or disappears. A small number of extremely young worker larvae will be chosen by the workers, who will then turn them into queen cells by demolishing nearby cells and constructing an extension to create a vertical cell. The chosen larvae are grown as queens and given additional supplies of a rich meal known as "royal jelly." Though a central location on the comb is typical, emergency cells can be found elsewhere in the brood nest. Usually, they are found in clusters of two or three cells.

A colony with swarming intentions raises swarm cells. The queen lays her eggs in unique cups that the bees construct for her. About two days after the swarm—half of the colony, including the old queen—has left, virgin queens begin to emerge. Queen bees generated from swarm cells are often larger than queens produced from emergency cells, and young larvae in swarm cells are exceptionally well fed with royal jelly. A colony will prepare 15 to 25 swarm cells, which are often located overhanging the bottom bar of the frames and around the margins of the brood combs.

Large, well-fed queen bees produced from supersedure cells are ready to take the place of an aging or ineffective queen in a colony. Similar to swarm cells, the queen lays her eggs in queen cell cups, which the bees prepare. The developing larvae are well-fed on royal jelly. Supersedure cells are often only produced in limited quantities by a colony at a time. Once the supersedure queen has mated and starts to lay she can co-exist with her mother in the same hive for a short period of time.

Artificial queen rearing: In order to artificially grow queen bees, you need to be able to instill in the bees the desire to produce a new queen for one of the three previously stated reasons. The swarm and supersedure impulse are the favored methods as they yield superior queens.

Conditions for rearing queens: Adequate factors are essential for a successful queen rearing process. Rearing queens at the incorrect time of year will yield inferior progeny. A moderate flow of nectar and a sufficient supply of pollen from three or more sources are ideal. In the event that pollen is scarce, premium pollen that has been gathered and preserved can be added. If nectar is insufficient, sugar syrup can be added as a supplement. The time is usually right to raise good queens if the swarming phase of the bee population is approaching. Some beekeepers feed colony-raising bees under any circumstances. In order to guarantee optimal nutrition for queen cells, this is strongly advised.

Supplementary feeding: Sugar and water can be added to queen-rearing hives at an equal volume ratio. Approximately one liter is fed every four to seven days. Choose a feeder that can be quickly replenished and operated with minimal difficulty. It is advised to use exterior bottle feeders or in-hive frame feeders; nevertheless, all syrup must to be consumed by the bees in fewer than three days to prevent fermentation and stomach issues. One way to supply pollen is to place a little amount of powdered pollen directly over the portion of the colony that houses the queen cells on a piece of board or tin. As the bees remove pollen, replenish the supply. To prevent the spread of bee diseases, it is essential to irradiate both purchased and self-collected pollen before transferring it to hives.

Breeding stock: The mother stock used to create producing queens with desirable features is known as breeding stock. The improvement of producing colonies is the main goal of queen rearing. To breed from unselected stock is to fail in this goal. You can either purchase or choose your breeders from respectable queen breeders or from your own apiary. Inbreeding will eventually cause the performance of the queens to decline if a single line of bees is continuously bred with only one or two breeders per generation. A closed breeding system should have as many breeder queens as feasible and at least 500 colonies. Accurate records should be the basis for selecting breeders for commercial queen breeding. Usually, the main criterion for selection is production. Brood rearing capacity, disease resistance, docility, swarming history, hygienic behaviour, and bee colony are additional criteria taken into account. The production of the colonies frequently reflects a number of these characteristics.

Equipment for queen rearing : Standard beekeeping equipment is used in most queen rearing systems, however some specialized equipment is needed. are designed to fit in the space occupied by a frame, and the plastic or wax cells—usually twenty per bar—are bonded to the bar using hot beeswax. To produce high-quality queen bees, bee colonies must be robust, disease-free, and pest-free. There should be at least two boxes full of bees and six to eight frames of brood in each of the two or three deck colonies that are used. Installing queen excluders is also necessary. For beginning cells and mating queens, three to five frame nucleus boxes will be required.

The Queen Rearing Process

Starting cells: Before a graft is inserted into a starter colony, it needs to be prepared. Because the grafted larvae in the two starting systems under discussion are "queen-less," meaning that, as in nature, they will want to replace their queen, the colony is tricked into accepting them under the emergency queen rearing drive.

1. Nucleus Box

A strong double hive should be turned to face the other way. With the nucleus entrance facing the same direction as the hive entrance, position a four-frame nucleus box behind the hive. The queen should be located in the double hive and placed to one side; the nucleus box should not include the queen. A feeder full of sugar syrup and one frame containing mostly pollen and a small amount of honey were placed into the nucleus box in that order. Two unsealed larvae frames and young bees from two other brood frames should be shaken into the nucleus box. If necessary, pollen or a protein supplement should be fed on top bars of the frames. To make room for shaking, remove the pollen frame; do not shake the queen into the nucleus. Put worker brood comb frames in place of the hive's frames that were removed. As soon as the nucleus colony is assembled, depart for at least five hours. A feeder full of sugar syrup and one frame containing mostly pollen and a small amount of honey were placed into the nucleus box in that order. Two unsealed larvae frames and young bees from two other brood frames should be shaken into the nucleus box. If necessary, pollen or a protein supplement should be fed on top bars of the frames. To make room for shaking, remove the pollen frame; do not shake the queen into the nucleus. Put worker brood comb frames in place of the hive's frames that were removed. As soon as the nucleus colony is assembled, depart for at least five hours. Refill the feeder by placing it against the super's wall; place additional pollen or a protein supplement on the upper bars. Insert the final three nucleus frames into the super. Put the 20-cell holding frame in the center of the super. With a maximum of 20 cells per feeding colony, additional bars of starting cells should be added to other hives that have been set up as cell feeders. Ensure that the brood chamber and the super are separated by a queen excluder.

2. Swathmore Swarm Box

A three- or four-frame nucleus box with a gauze bottom, risers ranging from 25 to 50 mm, and an entry closer in the closed position are needed for this procedure. Place one frame of pollen and one frame of honey within the "Swathmore" box. Bees from four frames of unopened brood from a robust colony shake into the box, being careful not to shake the queen into the box. Make sure the entrance is covered after shaking the bees into the box. Then, cover the box, store it in a cool, dark place for four to five hours. After that, graft up to forty queen cells and arrange the cell bars in a frame. To make it simpler to position the frame containing the grafted cells between the frames of honey and pollen, give the ground a strong bump before opening the box. Maybe some bees will get away. Overnight, keep the box in a cool, dark place. Check the cells by smoking under the lid and taking out the frame that holds the cell bars the next morning, or 16 to 20 hours after the cells were grafted. Now is not the time to bump the box. After that, the initiated queen cells are divided among the feeding colonies, with a maximum of 20 queen cells per feeder colony.

Cell Feeding Colonies

A queen right cell feeding or cell building hive is used to raise the queen cells to mature cell stage using the supersedure behavior of the colony, once they have been begun in the Swathmore or nucleus box. A robust colony will produce queen cells from seed. Make sure the queen cannot reach the queen cells. The following hives are appropriate for feeders or cell builders: 6–8 frames of disease-free brood, excluder between brood boxes, 6–12 months old, extremely robust, and well-maintained, with well-tempered bees covering the frames in both boxes. For every cell feeding colony, strong double hives with a plentiful supply of fresh pollen and nectar, inhabited by bees of all ages, are necessary.

Preparing cell feeding colonies - The day before it is needed, the cell feeding colony is assembled. The queen is discovered and placed under a queen excluder in the brood chamber.

In the center of the super, above the excluder, are two frames with unopened larvae that are 3-5 days old. To accommodate the frame containing the bar of grafted queen cells, a gap is left between the frames. A cell bar frame with one bar of the starting queen cells from the colony the cells were begun in is inserted between the brood frames in the super the day after grafting. To guarantee that every cell in the feeder colony gets enough food, no more than 20 queen cells should be added. Ten days after the cells were grafted, the cell bar is taken out of the feeder colony and divided into multiple queen cells for each mating nucleus. After being inserted into the mating nucleus, queens will come out of the cells one to two days later. Without the need for additional colony maintenance, a batch of cells can be produced once a cell feeder colony is established. If more batches of cells are to be added, they can be added five days after grafting the initial batch of twenty cells, or after the first batch is sealed.

Every seven days, cell feeding colonies that are used continuously should be adjusted. Below the excluder are the elder brood and the queen. In order to attract young nurse bees of an appropriate age (5 to 15 days old) to feed the queen cells, frames with unopened larvae 3–5 days old are moved above the excluder close to the queen cells. Support hives may need to be kept in the same apiary or in a different one if necessary in order to supply the cell feeder colonies with frames of bees, brood, pollen, and honey. Two days after being moved, the brood above the excluder should be examined, and any self-raised queen cells should be removed. If more than one batch of cells is being reared at once, you can maintain track by numbering the cell builder's lid and utilizing it as a record of the cells that are grown in. Mark the cell bar with your cell builder's number on the back of the bar ahead of the graft date (e.g., 5–10/10) when you check your graft and transfer it to your cell builder.

Grafting

Grafting is the process of transferring worker larvae into plastic cell cups when they are between 12 and 36 hours old, using a grafting instrument. A carved duck or goose feather, a metal grafting needle, a 00 point brush, a Chinese grafting tool, or a finely carved matchstick are a few of the grafting tools that can be employed. Both grafting out of dark combs and using a cool light or magnifying lamp (such as the x10 magnifying lamp used by crafters) will help you view the extremely little larvae. Utilizing a space away from the sun will prevent larvae from drying up and increase the likelihood of success. One important factor influencing the quality of the resultant queens is the age of the grafted larvae. The queens are better the younger the larvae. Choose the tiniest larvae you can find—ideally, ones that have just emerged out of the egg. An egg and a larva that are 0 to 24 hours old are the same length. Egg grafting is a very challenging process and is not advised. Strong colonies can be allowed to clean, polish, and heat the cells by adding plastic queen cell cups one day prior to grafting.

Grafting Method

To ensure that every larva is visible, place the frame holding the larvae at an angle of roughly 30 degrees on a support board. Cool light should be shining into the cells. Underneath the larva, place the grafting tool's nib or needle beneath the royal jelly. With the larva on top, scoop up some royal jelly. Next, center the nib's tip in the center of the queen cell cup's bottom, then roll the nib away from the larva so that it is now laying there in the center of the cell cup's base. If you have trouble seeing the larva, use a sharp knife to trim the cell walls at the foundation, and then remove the larva. If you believe that the larva you were transferring was harmed, try again with a fresh larva. Placing each larva in the center of the cell cup's bottom is crucial for their survival. To maintain your position on the bar while you graft into the cells, move your hand along the cells. When you are done grafting, quickly inspect your bar and regraft any that you believe to be damaged or overly large.

A tiny drop of royal jelly mixed with an equal amount of lukewarm water is often used by beekeepers to prime cell cups, but if royal jelly has been extracted and transferred with the graft, this step shouldn't be required. It is worthwhile to examine double grafting, which involves regrafting into cells that are one day old after the larvae are removed, solely for the purpose of producing breeding stock. Double grafting larvae has no benefit, according to research. When grafting several bars, keep the grafted bars moist by wrapping them in a towel until you're ready to insert them into starters and holders.

Checking Graft

After grafting, remove the cell bars and count the number of accepted cells the following morning (around 16–20 hours). Larvae in a pool of royal jelly and a wax extension on the cell cup's rim are features of accepted cells. After that, insert the bar with no more than 20 cells into the feeder hives. To form 20 cells, accepted cells may be transferred from one bar to another.

Transporting Queen Cells

Queen cells are sensitive, so handle them with caution. Handled incorrectly, cells can become dead or malformed queens. The tenth day following grafting is the safest time to transfer queen cells for distribution to colonies for hatching and mating. Sealed queen cells should always be kept hanging in the hive vertically. The fragile wing buds might be harmed if they are inverted. Transporting cells in a nucleus box with beeswax, honey, and pollen is the safest method. To avoid having your queen cells destroyed, make sure there isn't a queen present. In order to carry queen cells across long distances without the need for live bees to accompany them, portable incubators or insulated boxes have also been devised. To prevent vibration damage, the cells should be packed in warm sawdust, foam rubber, or another protective material.

Mating Queens

Young virgin queen bees will assault and destroy unhatched queen cells if they are permitted to emerge together. In order to avoid this, add queen cells to individual queenless colonies on the tenth day following grafting, or 1-2 days before emergence. Queen cells can be introduced directly into productive queenless colonies, although mated queens typically utilize a tiny colony known as a nucleus. Mating nuclei come in a variety of designs. Some are based on half length and shallow depth frame sizes and are specifically made for raising queens. Standard full-depth three- or four-frame nuclei work well for mating virgin queens in small-scale beekeeping operations. A lot of commercial beekeepers assemble nuclei over a split board or nucleus using standard material. This lessens the disturbance to the producing hive and makes it easier to introduce a new queen.

Beekeepers can also employ a typical single box with two or three frames of brood with bees and two frames of sealed honey and pollen, making up the difference with dry comb or foundation. Beekeepers can also kill elderly queens in weak or declining colonies. If hives are dispersed to allow queens to orientate during mating, this works well. To enable units to be locked in during assembly and for simple transportation to a different location at least two kilometers from the original hives, all nuclei and standard equipment should have ventilation in the lid or bottom board. Mature cells that are ten days old are positioned in the brood region between two combs, being careful not to compress the cells or move them from their upright position. After grafting, the virgin queen emerges on the eleventh or twelfth day, and she mates ten days later. After emerging from the cell, the young queens will begin to lay approximately twelve days later, if the weather has been cooperative. Up to 20 drones can mate with virgin queens. When virgins are ready to mate, there must be an enough number of adult drones available to ensure highly productive, long-lived queens. Drone brood hatches in 24 days, matures at 16 days, and experiences a 28-day drop in fertility until dying at the age of 55 to 60 days.

It is recommended to situate drone mother hives one to two kilometers away from the queen mating apiary in order to supply drones for mating with virgin queens. The virgin queens have nothing to do with drones. Drone mother colonies ought to be given the same consideration as a hive of cell feeders. Ten drone mother colonies are needed for each queen mating nuclei colony. Drone mother colonies require drones to be hatching in order to begin grafting queen cells. Approximately 700 drones will reach maturity within a whole drone brood frame. For mating queens, the design of the apiary is crucial. Mating is aided by a wind-sheltered location with useful landmarks, like as tiny shrubs; mating success is increased by spreading out the nuclei and making sure entrances are on various sides of a hive unit. Queen bees may return to the correct hive following mating flights with the help of painted hives, which can be colored differently, and painted shapes on the sides of nuclei.

Caging and introducing queens - Before introducing new queens, the old ones must be recognized and eliminated. It is possible to grow nuclei containing young laying queens into fruitful colonies or to remove the queen and replace it with a fresh, ripe queen cell. An alternative is to join the nucleus with a hive for re-queening. According to recent studies, when queens are brought into a new colony, older queens are more likely to be replaced than queens collected 28 days after emerging from the cell and kept in cages for mating. By papering on a nucleus hive, mailing cages, or Miller cages, new queens can be added to an existing colony.

Mailing cages - Once the queen comes, she can be introduced using the wooden or plastic mail-cages that are used to mail queens from the queen provider. After consuming the candy, the hive bees release the queen after 24 to 48 hours. Miller cages have a tiny entrance that is stocked with queen candy and are constructed of wire mesh and wood. They slide in and out of frames with ease and are reusable.

The process of papering a nucleus hive involves killing the old queen in the brood box, shaking out all the bees in the second box, putting two newspaper sheets between the two boxes, and then introducing the contents of the nucleus colony, balancing with old hive frames. Commercial beekeepers utilize a variation of this where they place a paper box containing a young queen on top of the old hive and let the bees reunite; in 90% of cases, the young queen survives and the old queen is killed.

Banking

Queen bees can be stored by the beekeeper until needed by banking mated queens. Wet weather, preserving queens for a big order, and introducing queens gradually over several weeks are the main causes of queen bee banking. It is possible to successfully bank queens for up to 21 days with little loss or harm to the queens while they are being stored. Either wooden bars with ten compartments or the mailing cages in which they arrive can be used to bank queens. Queen-right colonies above an excluder or queen-less hives have their queens banked in separate cages without escort bees. In order to keep a large number of young bees in the colony, strong, disease-free hives used for banking queens should be regularly supplied with frames of sealed brood and fed pollen and sugar syrup. Queen banking is a helpful management technique, but it is not risk-free. You should consult your equipment suppliers and queen before banking queens.

Documentation of appropriate timing is essential for effective queen bee rearing. Ten days after grafting, the cells must be divided, or else they will hatch and the new queens will annihilate one another. You also need to monitor the whereabouts of the cells. You will also need to know which breeder each queen comes from in order to evaluate them. Your record-keeping strategy could be as straightforward as making a note in a diary, but it must suit your needs. The following details are things to take into account: the date of grafting, the date of distribution, the date the queens emerged, the number of cells the bees started, the position of the feeder hives, the distribution of the laying queens, the numbers of the mating apiary sites, and the date the queen bees were captured. When it comes to record keeping, there are two options.

Queen Grafting Technique

Select a healthy colony with a genetically strong queen

↓

Remove the queen and make the colony queen less

↓

Select the larva of the age below (48hrs).
(Note: worker and queen larvae receive the same food for the first 48hrs)

↓

Transfer of larvae that were intended to be workers into queen cell cups. Cups may be dipped wax cups, moulded wax cups or plastic cups. These cups are affixed to bars and are positioned in frames

↓

These frames are inserted into queen less colony equipped with lots of nurse bees and lots of provisions such as honey (and/or syrup) and pollen (and/or Pollen particles)

↓

In the absence of queen, the workers feed the young larva with the secretion from their hypopharyngeal gland (royal jelly) for the development of new queen.

↓

On pupation, the cells are capped by the bees four or five days after grafting.

↓

Cover the queen cell with cell protectors keeping the newly emerged virgin queens confined, preventing them from being able to move about the colony and kill the other queens

↓

Virgin queens will emerge 15-17 days after the egg is laid 11-13 days after grafting. The average development time is 16days, but development is faster in warmer weather and slower in cooler temperature.

Fig. 1: Selection of a healthy colony with a genetically strong queen

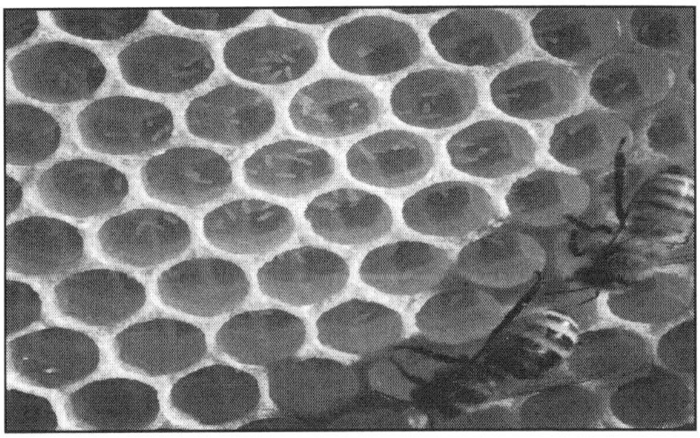

Fig. 2: Removal of the queen and the colony is made queen less.

Fig. 3: Selection of the larva age below (48hrs).

(Note: worker and queen larvae receive the same food for the first 48hrs)

Fig. 4: Transfer of larvae that were intended to be workers into queen cell cups.

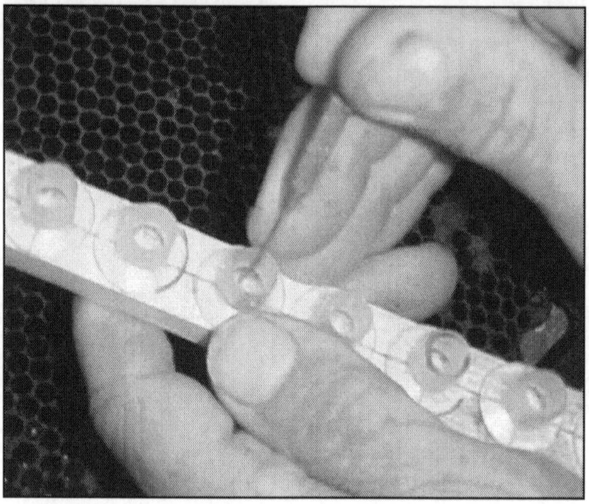

Fig. 5: Cups may be dipped wax cups, moulded wax cups or plastic cups. These cups are affixed to bars and are positioned in frames

Fig. 6: Transfer of larvae that were intended to be workers into queen cell cups. Cups may be dipped wax cups, moulded wax cups or plastic cups. These cups are affixed to bars and are positioned in frames

- These frames are inserted into queen less colony equipped with lots of nurse bees and lots of provisions such as honey (and/or syrup) and pollen (and/or Pollen particles)
- In the absence of queen, the workers feed the young larva with the secretion from their hypopharyngeal gland (royal jelly) for the development of new queen.
- On pupation, the cells are capped by the bees four or five days after grafting.

- Cover the queen cell with cell protectors keeping the newly emerged virgin queens confined, preventing them from being able to move about the colony and kill the other queens
- Virgin queens will emerge 15-17 days after the egg is laid 11-13 days after grafting). The average development time is 16ndays, but development is faster in warmer weather and slower in cooler temperature.

Chapter 11

Pests and Diseases of Honeybees

Honey, wax, and even bees are targeted by a variety of pests, predators, and foes. They consist of bee eater birds, lizards, frogs, mice, preying mantises, dragon flies, death's head moths, wax moths, wax beetles, ants, wasps, cockroaches, and so on.

1. Wax Moths

There are two different kinds of wax moths: *Achroia grisella*, the lesser wax moth, and *Galleria melonella*, the greater wax moth. These moths' waxy combs are consumed by their caterpillars. Within the hive, on the covered part of the combs, in the cracks and crevices of the walls, and on the bottom board of the hive, a female moth deposits 300–600 eggs in clusters. The smooth, round eggs range in color from pink to creamy white and measure between 0.4 and 0.5 mm. In summer, the eggs hatch in three to five days, while in winter, they do so in thirty days. The larvae begin eating on the honey, nectar, or pollen as soon as they hatch, if any are present. Subsequently, the larvae construct silken tunnels within the comb to obtain beeswax under shelter. The entire comb is covered in silken tunnels when the attack is intense. Older, darker combs are preferred by the larvae. Bees flee from such colonies. The larval stage lasts between thirty and forty days. The pupa is brownish in color and measures 14–16 mm in length. It eventually turns dark brown. Depending on the temperature, it takes one to eight weeks for the adult to emerge from the pupa.

Prevention and management: Since weak colonies are more vulnerable to wax moth attacks, management entails avoiding any circumstances that could cause the colony to weaken. These include a sufficient supply of food, defense against illnesses and pests, and protection of colonies from pesticide risks. Wax moth attacks can be avoided by routinely clearing the hives of debris and discarding old combs. For wax moths and their larvae, paradichlobenzene (PDB) is both poisonous and repulsive. You can sandwich two to four tablespoons of PDB between each super in a stock. But before using, the combs must be well aired out.

2. Wasps

Bees are attacked and killed by a variety of wasp species both in the wild and at the hive entrance. They include, among others, *Vespa magnifica, Velutina, Orientalis, Basilis, Tropica, Cincta,* and *Analis*. The wasps capture the approaching bees by hovering near the hive entrance. It is necessary to eradicate the wasp nests that are close by.

3. Wax beetle: *Platybolium alvearium.*

Nature of damage: On the lower board, among the debris, are both adults and grubs. They eat bits of the comb.

Prevention and control: On the lower board, among the debris, are both adults and grubs. They eat bits of the comb.

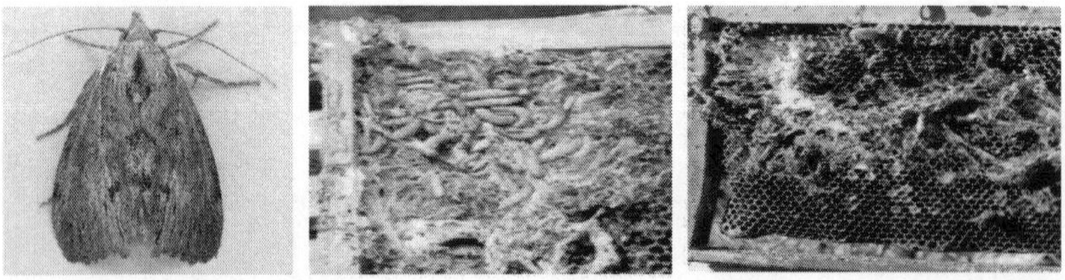

Greater wax moth, *Galleria melonella*

Lesser wax moth, *Achroia grisella.*

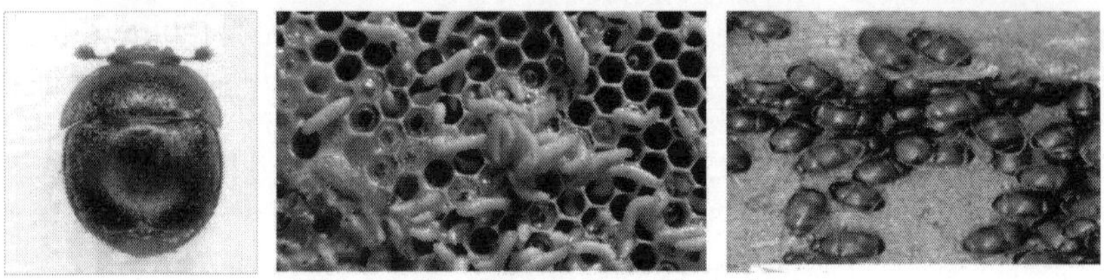

Wax beetle: *Platybolium alvearium.*

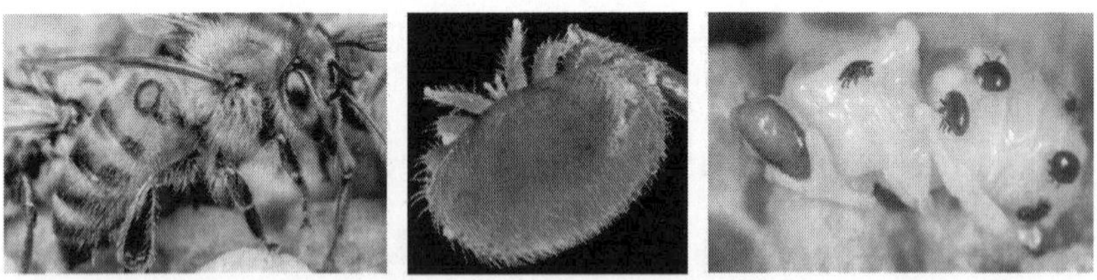

Varroa mite *Varroa jacobsoni*

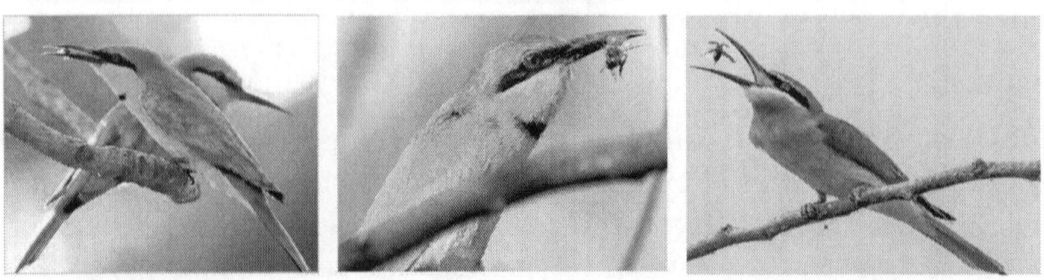

Bea eateres *Merops orientalis* *Merops phillippinus*

Pests of Honey Bees

4. Varroa mite: Varroa jacobsoni

Identification marks: the mature mite has a pin-head-sized size and is dark brown in color.

Nature of Damage : The mites cling to the bees and consume the intersegmental membrane in their abdomens. Additionally consume and destroy immature stages. The degree of deformity varies in the adult depending on the extent of damage.

Prevention and Control

- Application of sulphur dust @ 3 gm/lit on brood frames
- Fumigation with methanol, chlorobenzilate or formic acid

5. Crows

On cloudy days, king crows are likely to visit apiaries where they hunt bees.

6. Bee-eaters

In flight, the birds pose a powerful group of enemies for bees. With long wings and a tail that allow it to glide through the air with ease, the little green bee eater *Merops orientalis* is a stunning bird that is about the size of a sparrow and is always searching for flying bees. It swoops down and snatches the lady in its long, pointed beak as soon as it spots a flying bee. The behaviors of *M. supersiliosus* and *M. apiaster*, two other related species of bee eaters, are comparable.

Diseases of Honeybees

1. Nosema Disease

The protozoan *Nosema apis* is the source of this disease, since it attacks the stomach lining and causes dysentery. Its spores are dispersed by the infected bees' excrement, which worker bees consume while tidying the combs. During the winter, when bees hibernate, the disease primarily targets them and rapidly reduces their population. It is more prevalent in temperate nations than in tropical ones. When returning from field trips seeking nectar and pollen, infected bees that are robust enough to do so frequently succumb to exhaustion.

Control: Brood boxes and frame hives should only be sterilized after the queen and the bees have been moved to new foundation combs using glacial acetic acid fumes (soaked in cotton or rag at 120 ml/hive), ethylene oxide, or 40% formaline fumes. A novel medication, fumagillin, is very successful in reducing nosema illness when taken as a 2 mg dose in concentrated syrup.

2. Amoebic Disease

Malpighamoeba mellificae, a protozoan which infests the malpighian tubules, is another cause of it. The cysts that build up in the rectum or posterior end of the midgut infect the bees by ingestion. The cysts escape the enlarged, glassy malpighian tubules, enter the intestine, and are subsequently excluded along with the excrement. April and May mark the disease's springtime peak. Sometimes the virus is so bad that it wipes out the colony. Maintaining good personal hygiene and cleaning the equipment are the primary ways to contain the illness. The cysts are located on the bottom board and on the frames in feces.

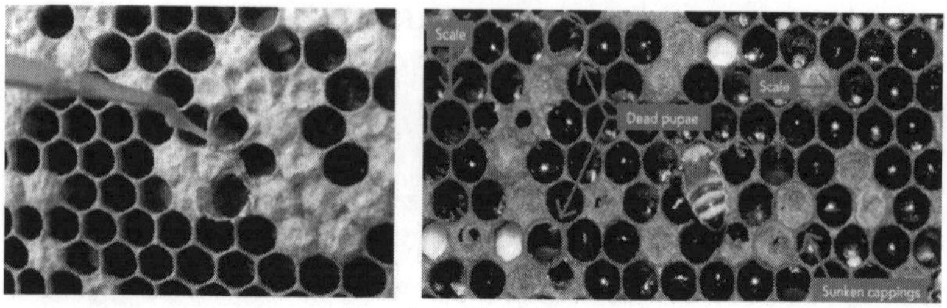

American Foul Brood

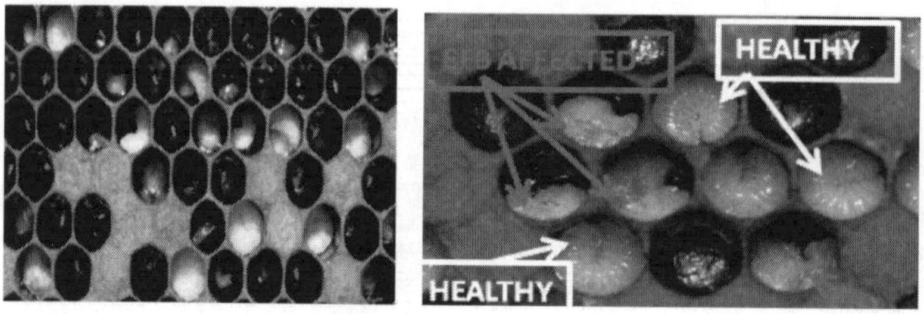

European Foul Brood

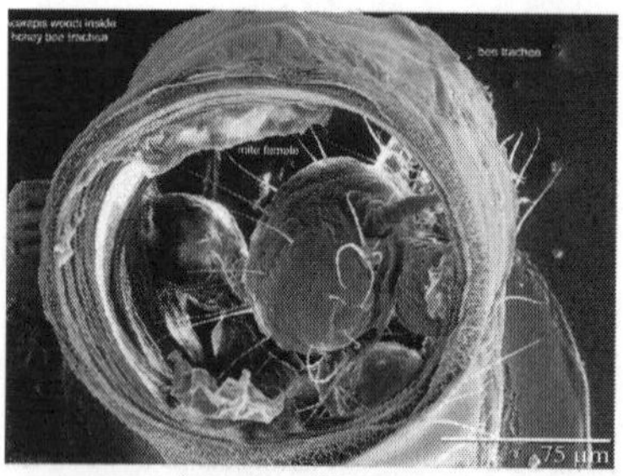

Acarine Disease

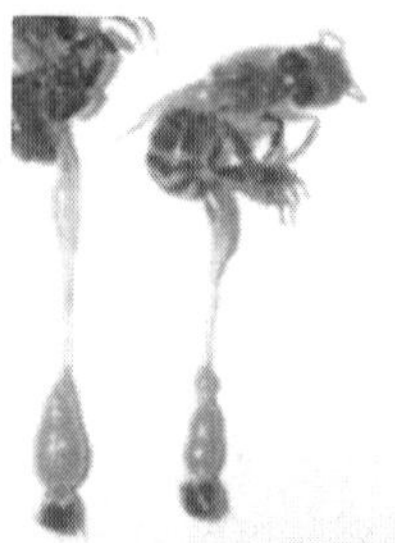

Nosema Disease

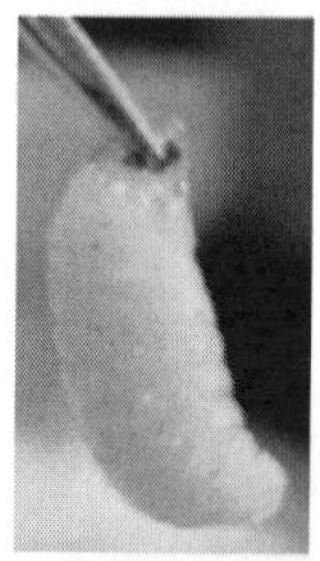

SAC Brood

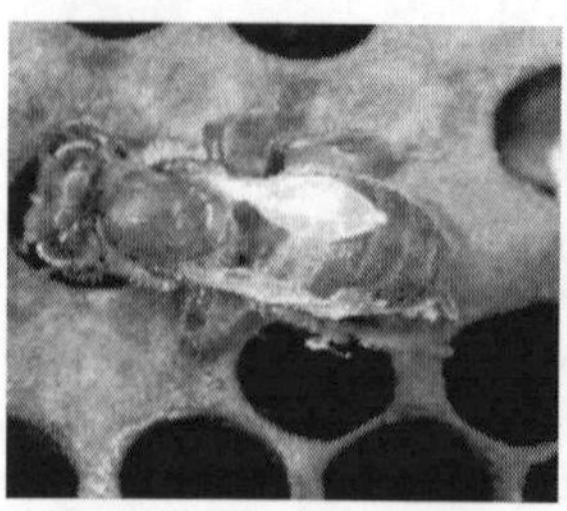

Chronic Bee Paralysis

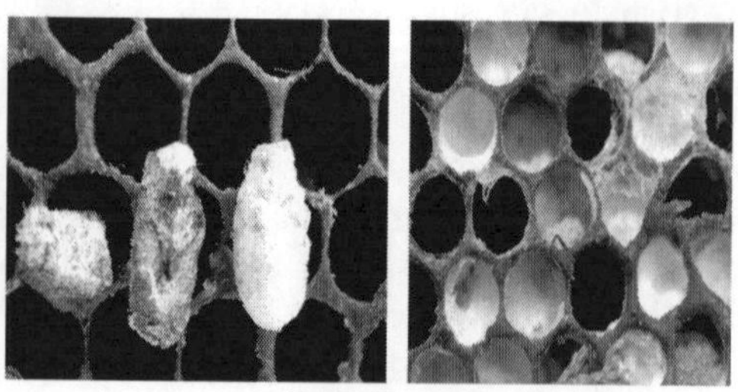

Chalk Brood

It needs to be removed with a scraper and then cleaned with 2% carbolic acid. The bees can be moved to boxes that have been fumigated with acetic acid for one week in the spring.

3. American Foulbrood

It is brought into the hive by bacteria called Baccillus larvae, which can enter through old beekeeping equipment, combs containing the excrement of infected bees, or spores in honey. Sunken and punctured cappings of sealed brood are the indications. Bees with patchy combs toss out the deceased brood when the cappings collapse. The deceased brood turns brownish-chocolate and then transforms into a sticky mess that smells bad. There's a foul stench emanating from the sick colony, kind of like rotting fish.

Prevention and Control: Applying sulphathiazole powder to the combs. Alternatively, afflicted colonies should be fed terramycin at a rate of 0.25-0.40 g/ 5 liters of sugar syrup, each week for six to seven weeks.

4. European Foulbrood

Melissococcus pluton, a nonspore-forming bacteria, is the cause. When young larvae consume infected brood food (royal jelly) within two days of their birth, they get sick and typically die before reaching adulthood. Even though they turn dark brown, the deceased larvae do not smell bad.

5. Chalkbrood (*Kalkbrut*)

Ascophaera apis, a fungus, is the responsible, turning larvae into clumps of chalk-white mycelium. The disease primarily affects the drone brood of honey bees, with worker brood being affected less frequently. Adults are unaffected. The consumed spores germinate, especially in the rear end of a cold larva's digestive track, and the hyphae infiltrate the tissues. If just one strain (+ or -) is present, the mummified larvae are white; however, if both strains are present, the fruiting bodies—which resemble tiny black dots on the larvae—form and the larvae turn dark grey or black.

6. Stonebrood (*Steinbrut*)

It is also caused by a fungus, *Aspergillus flavus* and less frequently *A. fumigatus* and other species of *Aspergillus*. Infection takes place by germination of the spores in alimentary canal and the adults may also be attacked. The infected larvae and pupae when dead turn into hard stone like mummies. The abdomens of dead adult bees also become mummified.

7. Sac Brood

The small ssRNA virus that causes sacbrood belongs to the Picornaviridae family.

8. Hai Sacbrood

It is also caused on by a virus that shares a lot of structural similarities with sac brood but is extremely potent against *A. cerana*.

9. Cute Bee Paralysis

Only adults are affected by this condition, which is brought on by the acute bee paralysis virus (ABPV), and they exhibit shaking, spread-eagled legs and wings. They occasionally also have a somewhat hairless appearance and, in certain cases, a lustrous black appearance. Even though the condition may have a high death rate, it may occasionally be minor or transitory. Adult infected bees typically die within a day after exhibiting the first signs of acute paralysis in two to four days.

10. Chronic Bee Paralysis

Unlike ABPV, which causes rapid death, this disease is caused by the chronic bee-paralysis virus (CBPV), which can cause an infected adult to live for several days. When adult bees contract CBPV, they might show two different symptoms. Adults with the first syndrome have an enlarged abdomen, which is brought on by the honey sac expanding when wet. Instead of flying, they crawl on the ground as their bodies and wings exhibit an unusual trembling action. At first, bees affected by the second condition are able to fly. They turn nearly hairless, dark, or nearly black, and glossy.

11. Acarine Disease

Acarapis woodi Rennie, a mite, is the cause of adult honey bees' Acarine sickness, often known as the Isle of Wight disease. In the tracheae of the bees, the mites reside, feed, and reproduce. They feed on blood and deposit their excrement in the trachea, causing damage to the linings when they pass through the spiracles on the first segment.

COLOUR PLATES

Prehistoric rock-painting of honey gathering, Arana cave, Spain, 8,000 BCE

Beekeepers of Ancient Egypt

Prehistoric bas-relief of a beekeeper harvesting honey, from Egyptian tomb, Pabasa, 7th-6th century BCE.

26th Egyptian Dynasty hieroglyph depicting a beekeeper with their hives of bees

Bee coin - Ephesus used the bee on its coins since it was a producer of honey

Ancient cylindrical beehives made of sun-dried mud from Israel

Wall painting of the tomb of Rekhmire (c. 1450 BCE) in Egypt, depicting horizontal beehives

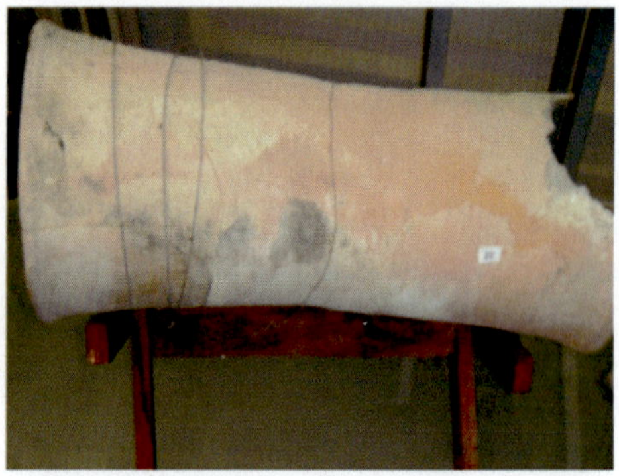

A traditional horizontal beehive from Greece (photo H. Harissis, Collection of A. Bikos in the Geoponic Institute, Athens)

Page 6

The Beekeepers, 1568, Painting by Pieter Bruegel the Elder

Prehistoric smoking pots from greece

Page 6

Stone bee hive at Malta, ancient Roman City

Part of the prehistoric apiary excavated at Tel-Rehov valley, Israel, 10th-9th century BCE

Bee house in Slovenia

Bee hives in Trees , Ethiopia

Page 6

Bamboo hive

Banana fiber hive

Straw bee hive

Skep

Pot hive with Stingless bees

Page 7

Barrel Hive with Circular frames

Wooden bee hive with straw

Pot hive

Page 7

ROCK BEE – *Apis dorsata*

INDIAN BEE – *Apis cerana indica*

Page 18

ITALIAN BEE – *Apis mellifera*

Page 18

Himalayan Giant Rock Bee, *Apis dorsata lamboriosa*

Himalayan Giant Rock Bee, *Apis dorsata lamboriosa hive*

STINGLESS BEE, *Trigonula irridipennis*

Page 21

STINGLESS BEE, *Trigonula irridipennis hive*

Page 21

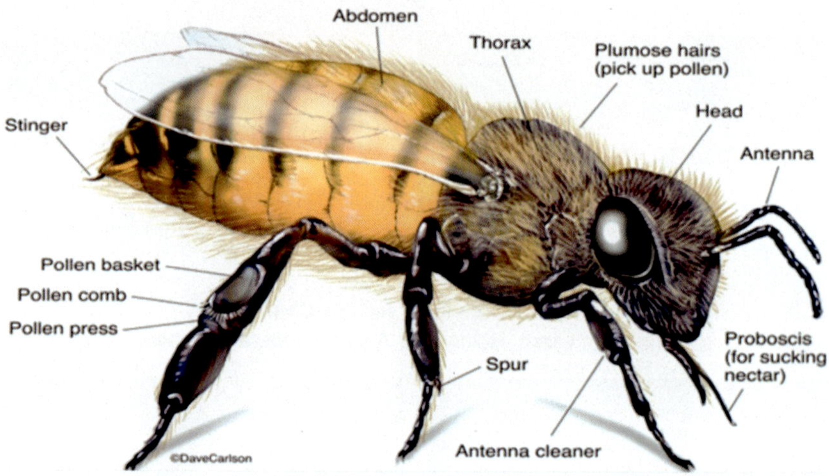

Honey Bee External Morphology

Honey Bee Mouth Parts – Chewing And Lapping Type

Page 25

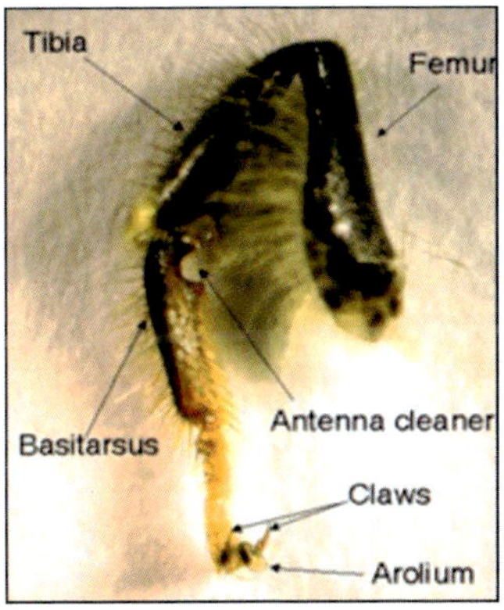

Front Leg-Antenna Cleaning

Middle Leg – Wax Picking

Hind Leg –pollen Basket

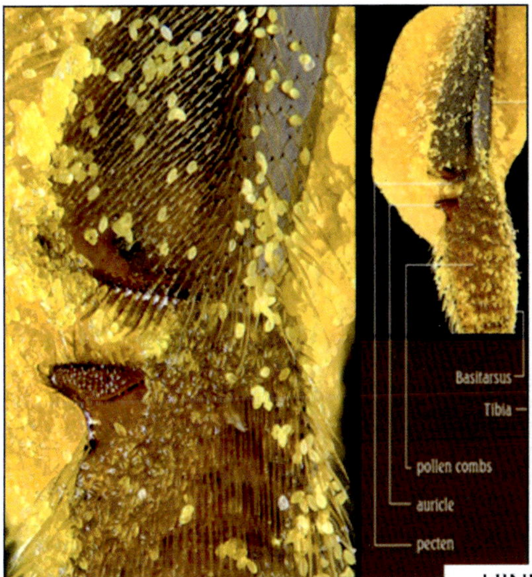

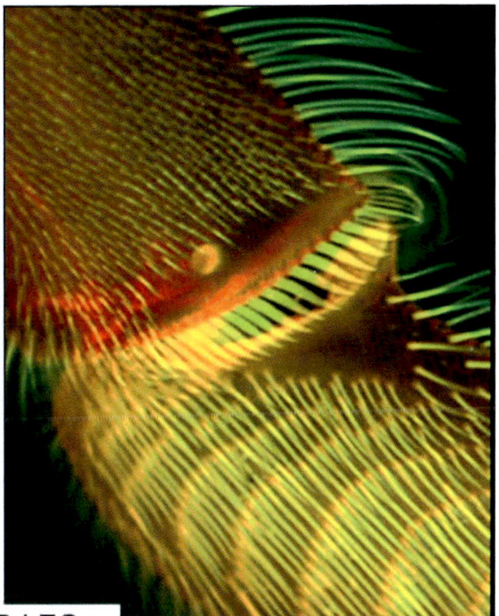

HIND LEG

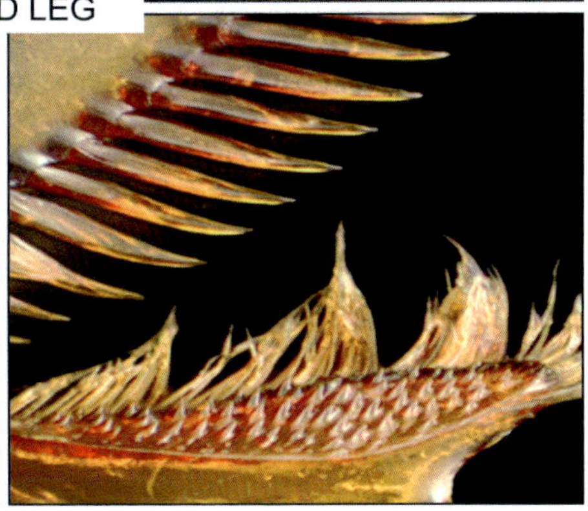

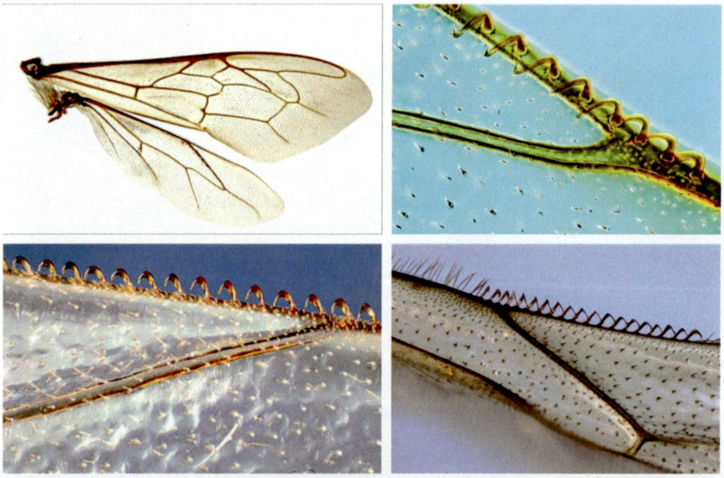

Honey Bee Wing Coupling - Hamulii

Page 29

Wax Secretion by Honey Bees

Honey Bees Constructing Bee Hive Using Wax

Page 31

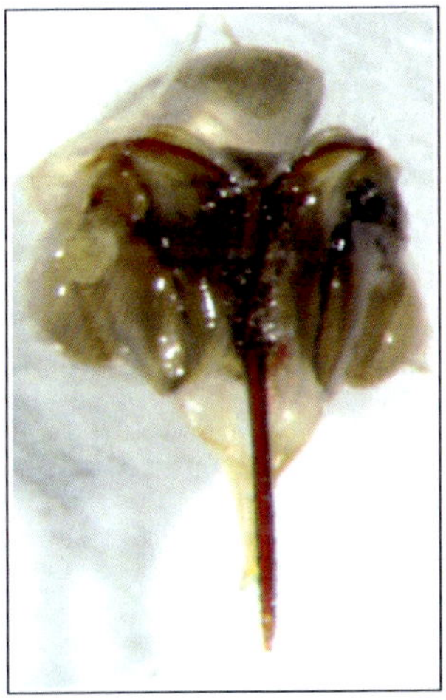

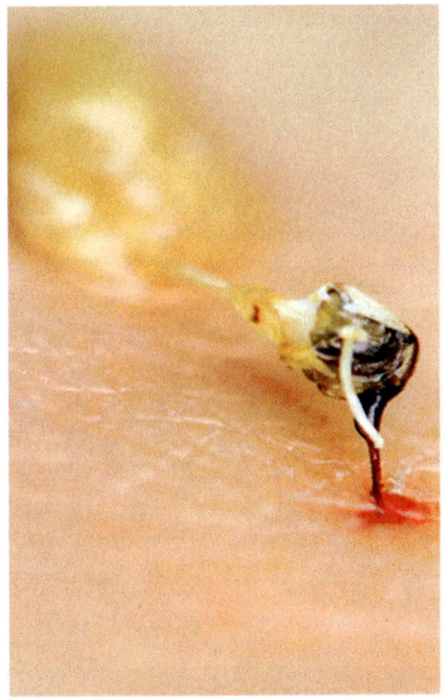

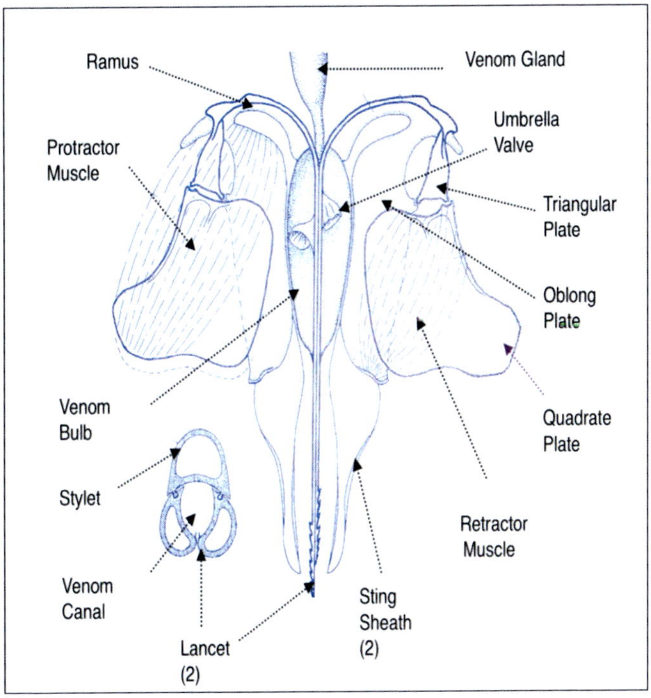

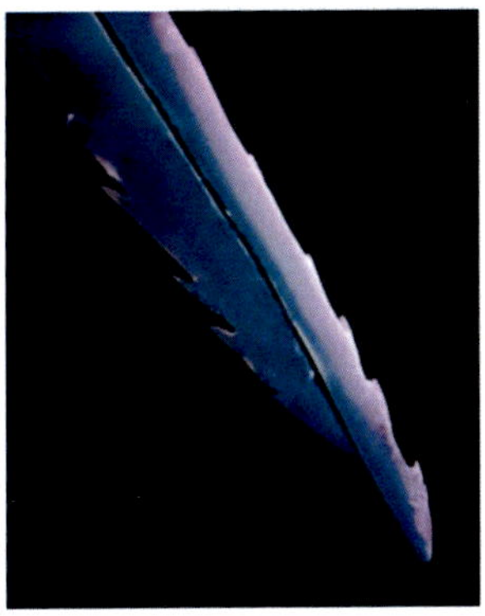

Honey Bee Sting

Page 32

Egg

Larvae

Pupae

Adult

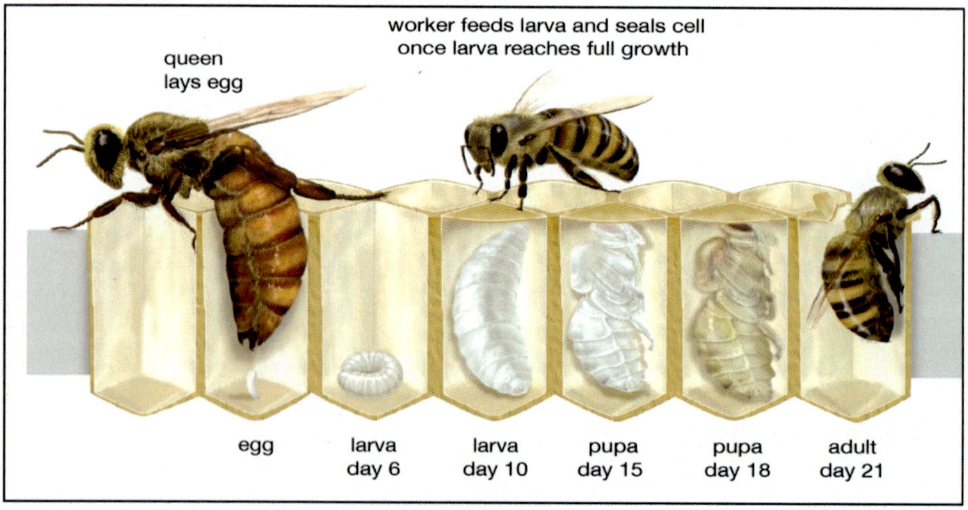

Honey Bee Biology

Page 45

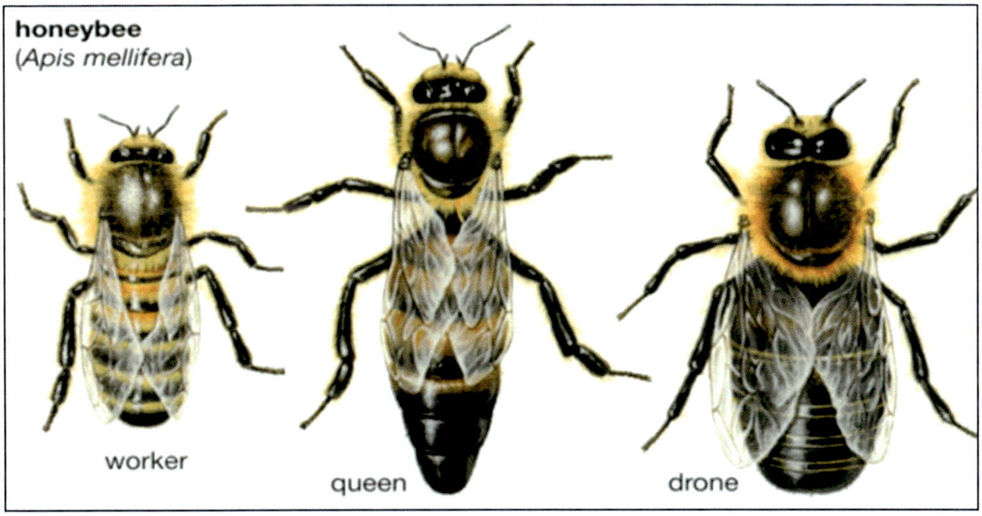

Honey Bee Colony

Page 48

CASHEW

CORIANDER

COCONUT

GOAT WEED

NIGER

SUNFLOWER

RAPESEED

RADISH

MUSTARD

TAMARIND

CUCUMBER

SUN HEMP

Page 52

WHITE SWEET CLOVER

KARANJ

CITRUS

DRUMSTICK

EUCALYPTUS

SESAME

LITCHI

SOAPNUT

BRINJAL

WATERMELON

ONION

SWEET POTATO

Page 54

Cosmos

Hibiscus

Golden rod

Aster

Tecoma stans

Zinnia

Edward rose

Coral creeper

Rangoon creeper

Page 55

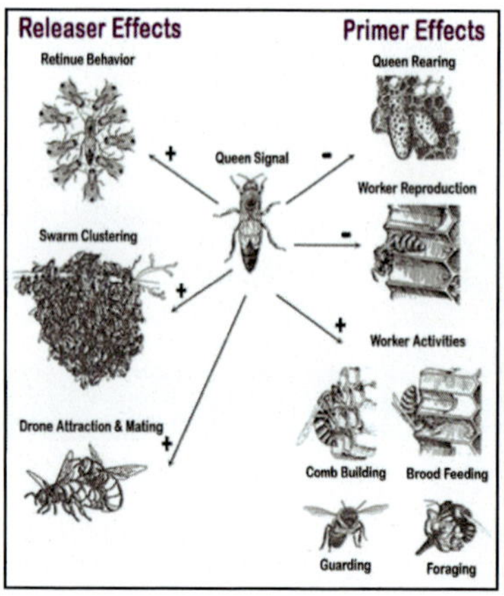

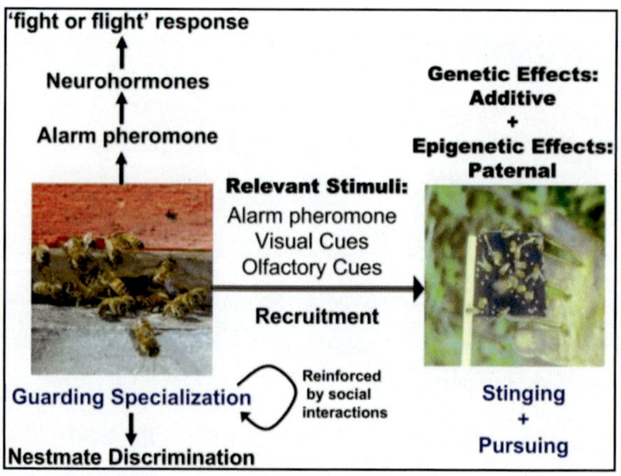

Page 59

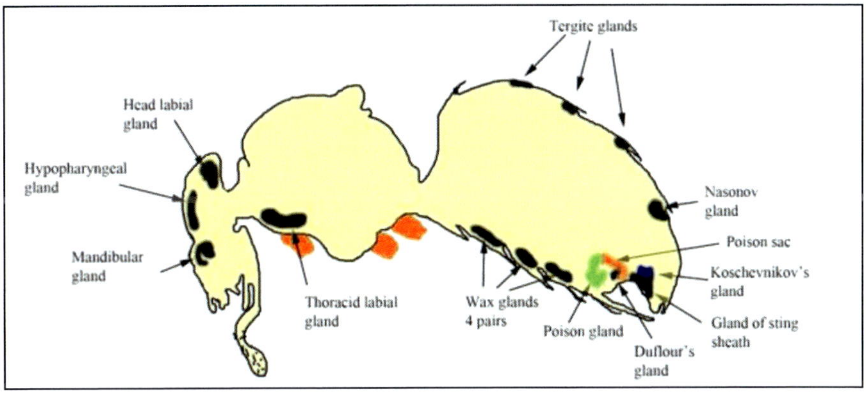

Page 62

Propolis

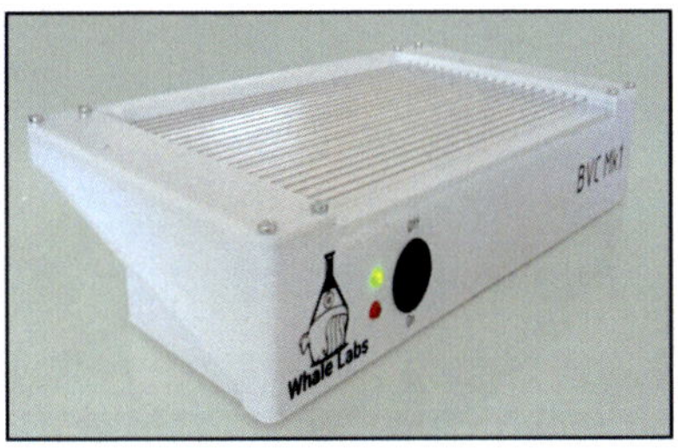

Bee Venom Collector

Bee Venom

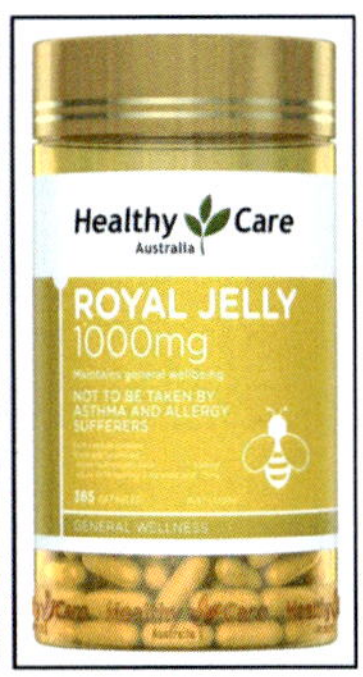

Royal Jelly

Page 83

Page 85

G. Param Jyothi of COH, VRGudem received 3rd position in AVISHKAAR- 2022, AP STATELEVEL INNOVATIVE Competitions organized by APCOST & AICTE ON 04 & 05TH March 2022 for preparing BEES WAX CRAYONS held at AdiKavi Nannayya University, Rajamahendravaram

Dr T. Janakiram, Hon'ble Vice Chancellor of Dr YSR Horticultural University congratulating the AVISHKAAR- 2022 award winners

Dr N.Emmanuel, Associate Professor (Ent) & ELP Beekeeping module in-charge Received EXCELLENCE IN INNOVATION AWARD – 2020 from DR B.V. VASANTH RAJ DAVID FOUNDATION, on 5th December 2020 at Le palace hotel, Chennai, for his Contribution in preparing multiple indigenous honey and beeswax value added products through ELP Programme

Page 105

Page 105

Page 106

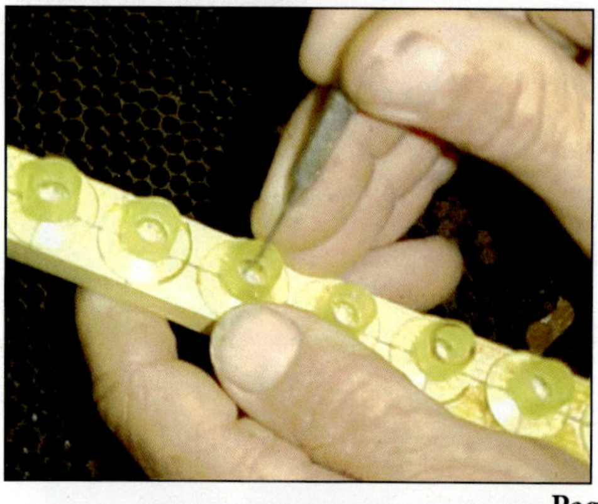

Page 106

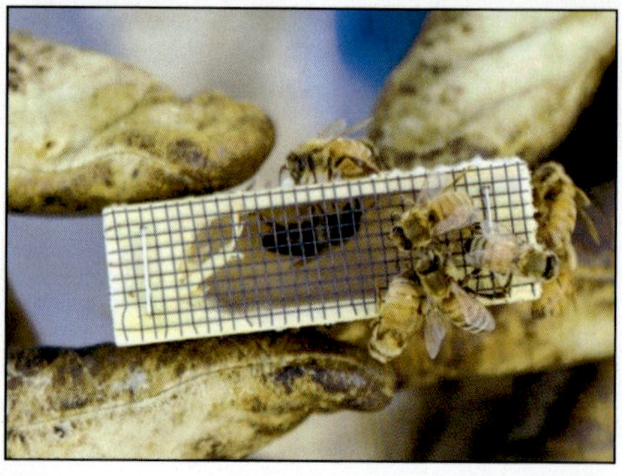

Page 107

Greater wax moth, *Galleria melonella*

Lesser wax moth, *Achroia grisella.*

Wax beetle: *Platybolium alvearium.*

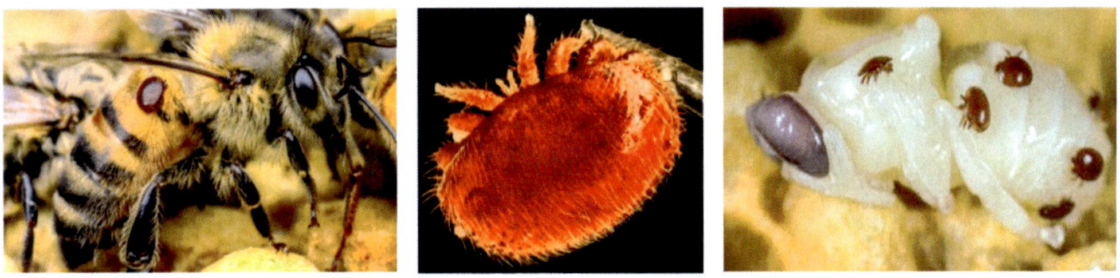

Varroa mite *Varroa jacobsoni*

Bea eateres

Merops orientalis

Merops phillippinus

American Foul Brood

European Foul Brood

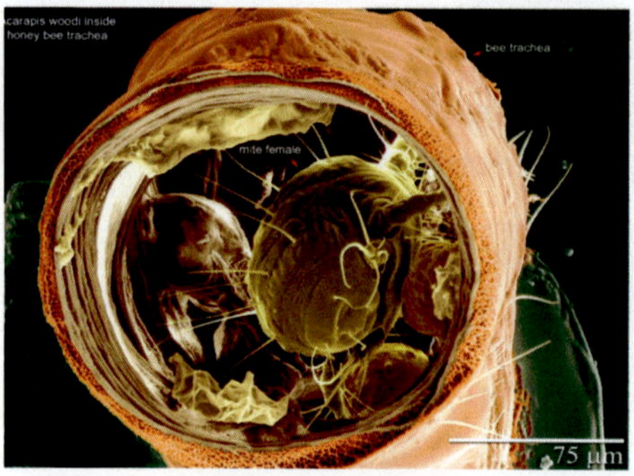

Acarine Disease

Nosema Disease

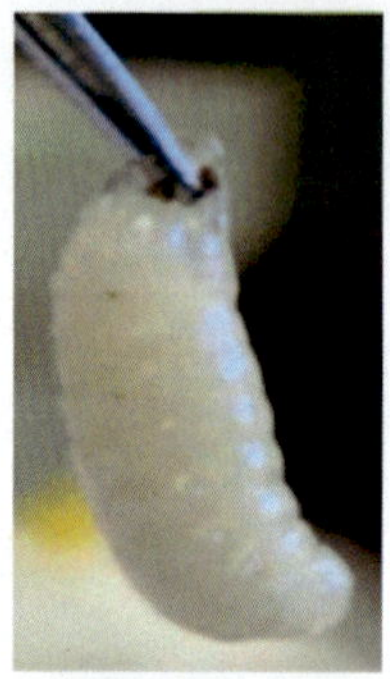
SAC Brood

Chronic Bee Paralysis

Chalk Brood

CHAPTER 12
Colony Collapse Disorder

Colony collapse disorder (CCD) is an unusual phenomena that happens when a colony of honey bees experiences a mass disappearance of worker bees, leaving behind a queen, an abundance of food, and a few nurse bees to tend to the remaining young bees. The disappearing disease, spring dwindle, May disease, autumn collapse, and fall dwindle disease are just a few of the names given to these sporadic disappearances throughout the history of apiculture. However, in early 2007, the syndrome was renamed colony collapse disorder due to a sharp increase in reports of disappearances of western honey bee (*Apis mellifera*) colonies in North America. A similar situation had been noted by beekeepers since 1998 across most of Europe, particularly in Southern and Western Europe; reports of a decline of more than 50% had reached the Northern Ireland Assembly. As it spread to various Asian and African nations, the phenomenon became increasingly widespread.

Because western honey bees are essential to the pollination of many agricultural crops worldwide, colony collapse disorder might result in large economic losses. The Food and Agriculture Organization (FAO) of the United Nations estimates that in 2005, the worth of all crops worldwide that were pollinated by honey bees approached US$200 billion. This information comes from the department of Agriculture and Consumer Protection. Bee shortages in the US have driven up the price farmers pay to rent bees for pollination services by as much as 20%. Bee population declines, however, predate CCD by several decades: since 1961, the US managed hive business has been steadily contracting.

On the other hand, since 1975, the global bee population has been growing significantly due to honey production; China has contributed the majority of this expansion. The years 1991–1999 had the lowest rise in the production of honey worldwide because of the economic crisis that followed the fall of communism in the former Soviet area of influence. Despite CCD, production has grown by 50% more in 2020 compared to 2000, more than doubling the rate of growth in earlier decades.There have been several theories put up on the causes of CCD, however none of them have been widely accepted by scientists. Pesticides; infections with different diseases, particularly those carried by Acarapis and Varroa mites; malnourishment; genetic factors; immunodeficiencies; habitat loss; altered beekeeping techniques; or a combination of these are some of the hypothesized causes. Though many collapsing apiaries show no trace of neonicotinoids, much conjecture has surrounding the contributions of this family of chemicals to CCD.

Signs and Symptoms

CCD is not the same as colony decline, which can be brought on by a number of problems, including nutrition, varroa mite infection, queen health, and other illnesses. When it is noticed that there are not many adult bees physically present in the colony after a collapse, CCD is considered. There are very few, if any, dead bees found in or around the hive, as if it had just been abandoned, in contrast to other acute sources of die-off like pesticide exposure. All of these circumstances typically occur at the same time in a colony that has fallen due to CCD.

- Presence of capped brood in abandoned colonies. Bees normally do not abandon a hive until the capped brood has all hatched.
- Presence of food stores, both honey and bee pollen:
    - that other bees do not rob immediately
    - with significantly delayed attacks by hive pests such as wax moths and small hive beetles
- Presence of the queen bee. If the queen is not present, the hive likely died because it was queenless, which is not considered CCD.
- No dead honey bee bodies present
- Precursor symptoms that may arise before the final colony collapse include:
- Inability to maintain current brood due to low workforce
- Colony includes mostly young adult bees
- Bees are reluctant to consume provided feed, such as sugar syrup and protein supplement.

A number of physio-pathological characteristics may function as biomarkers for colony health and predict CCD status prior to any clinical manifestation of colony collapse disorder. Collapsing colony bees typically have soft feces, partially filled rectums, rectal enteroliths, or rectal stones, and iridescence in the Malpighian tubules. While rectal enteroliths indicate a breakdown of excretory physiology, which may further lead to constipation and poor osmoregulation in CCD bees, a faulty rectum implies nutritional disturbance or water imbalance. It has been established that these characteristics do not correlate with age and manifest to varying degrees in four different age groups of bees: newly emerging bees, nurse bees, non-pollen foragers, and pollen foragers.

Furthermore, there are genetic markers in the gut that point to honey bees' vulnerability to CCD. It has been established that 65 distinct RNA transcripts may serve as indicators of CCD status. When compared to healthy bees, the genetic expression of these transcripts was either elevated or downregulated based on the genes. Through the use of qPCR and microarray analysis, it was possible to identify the quantity of unique ribosomal RNA (rRNA) fragments with poly(A)-rich 3′ tails in the guts of CCD bees. Based on available data, it appears that these poly(A)-rRNA sequences function as degradation intermediates to support rRNA's enzymatic activity and protein folding. Moreover, poly(A)-rRNA expression and the presence of the Israeli acute paralysis virus and deformed wing virus are genetic markers for the onset of CCD.

United States

According to data from the National Agricultural Statistics Service (NASS), there were 2.44 million honey-producing hives in the US in February 2008, down from 4.5 million in 1980 and 5.9 million in 1947. However, it should be noted that these figures may not accurately reflect the total number of managed hives in the country, as they do not include hives managed by beekeepers who own fewer than five hives or thousands of hives managed for pollination contracts. The practice of counting some hives more than once—hives that are transported to multiple states to produce honey are tallied in each state's total and added in overall counts—might compensate for this underrepresentation.

At least 24 states in the US had reported at least one incidence of CCD in 2007. A 2007 survey of 384 participating beekeepers from 13 states revealed that 23.8% of them satisfied the CCD criteria, which is that 50% or more of their dead colonies were discovered to be bee-free or with very few dead bees in the hive or apiary. Between 2006 and 2007, the total loss of all colonies suffered by beekeepers with CCD was 45%, while that of beekeepers without the disease was 25%.

Following the US description of CCD in 2007, the yearly winter colony losses rose from 15% to 30%. Between 2014 and 2017, the percentage of such losses dropped to 24%, and CCD symptoms were not as frequently linked to bee losses. Since the discovery of CCD, the number of honey bee colonies in the US has either stayed constant or increased, despite the fact that CCD has increased hive losses.

According to NASS, the total number of hives in the United States in 2017 varied from 2.63 to 2.99 million for businesses with more than five colonies to 35–43 thousand hives for those with fewer than five colonies over the course of the year. Operations with more than five colonies experienced a loss of 77.8 thousand hives (2.6–3.0%) with CCD symptoms in the same year, while operations with fewer than five colonies experienced a loss of 6 thousand hives (14–17%).

Europe

The European Food Safety Authority (EFSA) reports that in 2007, there were 274,000 hives in the United Kingdom, 1,091,630 in Italy, and 1,283,810 in France. 2008 saw a 30% decline in the bee population in the United Kingdom, according to the British Beekeepers Association, and a 40–50% mortality rate in Italy, according to an EFSA research. EFSA authorities note that the results are not entirely trustworthy, though, because prior to the bees' deaths, there was no standardization in the methods used by various nations to gather data on their bee populations. Reports at the time (2008) attributed the high death rate to the varroa mite, two seasons of exceptionally rainy summers in Europe, and certain chemicals.

Anecdotally, it is extremely variable, according to Tim Lovett, president of the British Beekeepers' Association in 2009. Some beekeepers are said to have lost nearly a third of their hives, while others have lost none at all. The chairman of the London Beekeepers' Association, John Chapple, estimated that between a fifth and a quarter of his 150 members suffered losses. "There are still a lot of mysterious disappearances; we are no nearer to knowing what is causing them." The National Bee Unit of the government persisted in disputing the presence of CCD in Britain, blaming the significant losses on the varroa mite and wet summers that prevent bees from searching for food.

British Beekeepers' Association member David Aston said in 2010 that "we Sstill do not believe CCD (which is now better defined) is a cause of colony losses in the UK, however we are continuing to experience colony losses, many if not most of which can be explained." He believes that current research provides "further evidence to the evolving picture that there are complex interactions taking place between a number of factors, pathogens, environmental, beekeeping practices and other stressors, which are causing honey bee losses described as CCD in the U.S.

Scottish beekeepers also reported losses between 2007 and 2009. In the winter of 2009–2010, Andrew Scarlett, a bee farmer and honey packer from Perthshire, lost 80% of his 1,200 colonies. He blamed the losses on a severe bacterial infection that developed swiftly due to a dearth of bee inspectors and persistently bad weather that kept honey bees from accumulating enough reserves of pollen and nectar. There was no scientific proof in Germany, the country where some of the first reports of CCD in Europe surfaced and where the German National Association of Beekeepers said that 40% of their honey bee colonies perished. Early in May 2007, the German media claimed that there didn't appear to be any confirmed instances of CCD in Germany.

The first occurrence of CCD outside of the United States was a colony loss incident in Switzerland that was surveyed in October 2009 by the Swiss Bee Research Centre, according to a case study published in 2012. Although the incedent met the parameters proposed by previous US hive surveys, examination of pathogen-related laboratory testing revealed no difference in pathogen loads or levels of co-infection between CCD and non-CCD hives. The case study also points out that at the time of collapse, the colonies did not exhibit harmful concentrations of Varroa destructor or Nosema spp., but it underlines that infections could not be completely ruled out as being involved. Approximately 50% of the bee population was estimated to have perished throughout the winter by the Swiss government at the end of May 2012. It was believed that the parasite Varroa destructor was the primary cause of the decline.

Asia

A three-year survey conducted in China between 2010 and 2013 revealed an average decrease of 10.1% of colony size. Queen issues and comb renewal were found to be important risk factors.

Possible Causes for CCD

Although the exact causes of CCD are still unknown, a wide range of factors, including pesticides, mites, fungi, beekeeping practices (like using antibiotics or moving beehives over long distances), malnutrition, poor quality queens, starvation, other infections, and immunodeficiencies, are currently being investigated. The current agreement in science is that multiple elements working together can either additively or synergistically cause CCD rather than any one factor alone.

The Colony Collapse Disorder Working Group (CCDWG) was founded in 2006 and is largely housed at Pennsylvania State University. Their initial report identified certain trends but did not come to any firm conclusions. According to a survey conducted early in 2007, the majority of hobbyist beekeepers thought that starvation was the main cause of death in their colonies, while the majority of commercial beekeepers thought that invertebrate pests, such as small hive beetles, honey bee tracheal mites, and Varroa mites, were the main cause of colony mortality. Similar to this, a June 2007 scholarly review considered a number of theories and potential contributory factors but could not settle the issue.

A CCD Action Plan, published in July 2007 by the US Department of Agriculture (USDA), detailed a four-pronged approach to addressing CCD: sample analysis and surveying; hypothesis-driven research; mitigation and preventive action. 2009 saw the release of the U.S. Colony Collapse Disorder Steering Committee's inaugural annual report. It was suggested that a variety of factors working together could be the cause of CCD.

A thorough descriptive analysis conducted in the same year by the CCD Working Group came to the following conclusion: "No single factor was found with enough consistency to suggest one causal agent out of the 61 variables quantified (including adult bee physiology, pathogen loads, and pesticide levels)." The pathogen loads and co-infection rates of CCD bee colonies were higher than those of control populations, indicating a possible increased pathogen exposure or weakened defenses in CCD bees."

November 2010 saw the publication of the second annual Steering Committee report. Despite the fact that numerous correlations, such as those with pesticides, parasites, and infections, had been found during the course of the investigation, the group stated that "it is becoming increasingly clear that no single factor alone is responsible for [CCD]". According to their findings, there were no harmful concentrations of parasitic Varroa mites or the parasite Nosema at the time of collapse. They did discover a link between the sublethal effects of some pesticides and CCD. Two prevalent miticides in particular—fluvalinate and coumaphos—that are approved for use by beekeepers to control varroa mites are among these pesticides. Research has also revealed the sublethal impacts of fungicides and neonicotinoids, insecticides that may weaken bees' defenses against viruses that infect them.

A review conducted in 2015 looked at 170 papers on colony collapse disorder and bee stressors, including as agrochemicals, viruses, diminishing biodiversity, and climate change, among other things. The investigation found that "a strong argument can be made that it is the interaction among parasites, pesticides, and diet that lies at the heart of current bee health problems." Moreover:

Pathogens and Immunodeficiency Teories

Early investigators observed that the spread route behaves like a communicative illness; on the other hand, there was a belief that the illness might have an immunosuppressive mechanism, possibly connected to "stress" that compromises immunity. In particular, a 2007 study from Pennsylvania State University found that "the magnitude of detected infectious agents in the adult bees suggests some type of immunosuppression". Initially, these researchers proposed a link between CCD and infestations of Varroa destructor mites, hypothesizing that a combination of these bee mites, deformed wing virus (which the mites spread), and bacteria cooperate to reduce immunity, which could be a contributing factor to CCD. Beekeepers face a variety of problems that have been introduced within the last 20 years in the continental United States, including kleptoparasites like small hive beetles (*Aethina tumida*), bacterial, viral, and fungus diseases, and parasites like varroa mites (*Varroa destructor*) and honey bee tracheal mites (*Acarapis woodi).*

In a bee yard, when a colony is dying for any reason and there are healthy colonies nearby, the healthier colonies will often enter the dying colony and take its resources for themselves. A trend of healthy colonies getting sick while near a dying colony could indicate to an observer that a contagious disease is present if the dying colony's food supply was tainted (by natural or man-made toxins). But in most CCD situations, supplies of dying colonies are not pilfered, indicating that toxins do not spread by pilfering, imitating a disease.

The following observations provided additional evidence that CCD is an infectious disease: the CCD Working Group report from 2010 suggested that CCD-exhibiting hives tended to occur in close proximity to one another within apiaries, and that the hives of colonies that had died from CCD could only be reused with a healthy colony if they had first been treated with radiation that destroys DNA.

Varroa *Mites*

A 2007 report claims that the mite Varroa destructor is still the most damaging honey bee killer in the world. This is partly because of the viruses it carries, which include acute bee paralysis virus and deformed wing virus, both of which have been linked to CCD. Bees that suffer from Varroa mite infestation also typically have weakened immune systems. Enesto Guzman, a Canadian entomologist at the University of Guelph, examined 413 Ontario bee colonies between 2007 and 2008.

It has been shown that the presence of Varroa mites within colonies prior to winter weakens the bees' immune systems and introduces viruses that cause the colonies to die throughout the winter. In 85% of cases, the Varroa mite was shown to be the cause of the approximately 27% of hives that did not survive the winter. Additionally, varroa mites hinder the queen's capacity to procreate, which is harmful to the hive's survival. As a result, although not all dying colonies contain these mites, Varroa mites have been suggested as a potential cause of CCD.

A parasite mite called Varroa destructor takes over beehives and feeds on honey bee hemolymph. Depending on the stage of their life cycle, varroa mites parasitize all varieties of honey bees, including worker, nurse, and larval bees. Varroa like to connect to nursing bees during the phoretic stage because this increases their fitness before entering the reproductive stage. Subsequently, the mites consume larvae during their reproductive phase, and heightened fitness results in a rise in mite fecundity, or the quantity of female progeny. Varroa is one of the main hazards to colonies since they can feed on any kind of honey bee, especially during the winter.

In 2020, a team of researchers declared that they were in the initial phases of field testing a bacterium that included specially altered plasmids that not only inhibited the infection of deformed wing virus but also successfully decreased the survival rate of Varroa mites.

Israeli Acute Paralysis Virus

Once identified as the cause of CCD, the Israeli acute paralysis virus (IAPV) was found in Israel in 2004. Its original location is unclear, however it was called for the location where it was initially discovered. Results of a large-scale statistical RNA sequencing investigation of colonies affected and unaffected were released in September 2007. To find out if diseases were present, RNA from every organism in a colony was sequenced and compared with sequence databases. Numerous pathogens were detected in all the colonies, but only the IAPV virus had a noteworthy correlation with CCD: the virus was detected in 25 out of the 30 examined CCD colonies and in only one of the 21 tested non-CCD colonies.

A 2009 study discovered that all CCD-affected bees shared an indicator for reduced protein production, a pattern that was determined to be compatible with IAPV infection. Like the IAPV, it is hypothesized that Dicistroviridae degrade ribosomes, which are in charge of producing proteins in cells. This diminished ribosomal activity is thought to weaken bees and increase their susceptibility to potentially fatal stimuli.

Nosema

Nosema ceranae infections in western honey bee hives eradicate the infection in eight days, suggesting that *N. ceranae* may be the cause of CCD. Although a 2009 investigation of US CCD-affected bee populations indicated that only around half of the examined colonies, both in CCD and control populations, were infected with *N. ceranae,* the research team claims to have ruled out many other possible causes.

Fungagillin is the main antifungal agent used against Nosema; it has been utilized in a German research study to lessen the impact of the microsporidian and is listed by the CCDWG as a potential treatment. Higes further asserts that fumagillin has been effective in curing colonies. While describing these results as promising, one analysis issued a warning, stating that "*N. ceranae* may not be to blame for all cases of colony collapse". Although this fungus has been observed in several parts of Europe, no connection to CCD has yet been found.

A few hives in California were found to have *N. ceranae* in 2007. However, the researcher did not think that this proved there was a clear connection between CCD and the condition; "We don't want to give anybody the impression that this thing has been solved." Similar words have been spoken by a USDA bee scientist: "*Nosema ceranae* is not the only cause; it is just one possible factor. The fungus has previously been observed, occasionally in colonies that were in good health.

Using PCR of the 16S gene, *N. ceranae* has been found in honey bees from multiple states. 49 counties in New York have been found to have *N. ceranae*. Of the 1,200 honey bee samples that were collected, 528 (44%) tested positive for Nosema. Of those samples, 371 had PCR analysis that revealed 96% contained *N. ceranae*, 3% included both *N. ceranae* and *N. apis*, and 1% contained N. apis alone. Healthy bees are more likely to contract *N. ceranae*, which may indicate a connection to CCD, when they are given pollen laced with fungicides, insecticides, and other agrochemicals, such as imidacloprid.

Viral and Fungal Combination

In October 2010, a group of scientists from the University of Montana and Montana State University, led by Jerry Bromenshenk, and collaborating with the US Army's Edgewood Chemical Biological Center, published a paper stating that every dead colony the group examined contained the fungus *Nosema ceranae* and a novel DNA virus known as invertebrate iridescent virus type 6 (IIV-6). According to their research, the virus and N. ceranae together were always 100% lethal, but neither agent by itself appeared to be harmful. The public was informed about the study through a front-page piece published in The New York Times.

Pesticides

The USDA claims that pesticides might be a factor in CCD. For a sometime, scientists have been worried that pesticides, and perhaps even some fungicides, could have sub lethal effects on bees, affecting their growth and behavior rather than killing them outright. "Pesticides alone have not shown they are the cause of CCD," stated Maryann Frazier. We think that a number of variables are at play, potentially including pesticides, pathogens, and mites."

It is particularly challenging to assess the impacts of pesticides to CCD for a number of reasons. First, testing for every potential pesticide at the same time is challenging due to the wide range of pesticides used in the many areas reporting CCD. Second, a lot of commercial beekeeping operations are mobile, moving hives over great distances over the season, which could expose the colonies to various chemicals depending on where they are kept.

Third, there is no way to link the onset of symptoms to the precise moment of pesticide exposure because the bees themselves store pollen and honey for long periods of time. This means that it could take weeks or even months before contaminated food is supplied to the colony. The vulnerability of honey bees to pesticides may have been explained in 2010 by a genome sequencing project. The number of genes encoding detoxification enzymes, such as carboxylesterases, glutathione-S-transferases, and cytochrome P450 monooxygenases (P450s), is lacking in its genome.

Neonicotinoids

Neonicotinoids, a class of insecticides, have drawn special attention. Systemic insecticides called neonicotinoids are usually applied as seed treatments to minimize the need for foliar pesticides as plants develop. The three principal substances used in agriculture are thiamethoxam, clothianidin, and imidacloprid. Field experiments that fail to demonstrate effects on real bee populations have frequently been conducted after potentially hazardous effects evaluated in the lab. Regulators have limited the use of neonicotinoids in Europe and other regions due to worries about the health of bees, despite gaps in the scientific data.

Evidence of Risk to Bees

Neonicotinoid pesticide use surged in the United States following 2005, coinciding with an increase in bee mortality. The majority of corn (maize) planted in the United States is cultivated from neonicotinoids-treated seeds. Despite the fact that maize is pollinated by the breeze, honey bees may come into contact with pesticide residues in the nectar and pollen when they are feeding on the plants. Additionally, honey bees may unwittingly come into contact with nicotinoids when foraging on wild plants. Although beekeeper surveys have been the main source of information used to assess potential chemical roles in CCD, direct testing of samples from impacted colonies appears to be necessary.

Laboratory Studies

Neonicotinoids may obstruct bees' innate sense of direction, leading to disorientation and making it difficult for them to return to the hive. Neonicotinoids may have an impact on bees' short- and long-term memory, leading to these deficits. In laboratory and field trials, imidacloprid at sublethal levels reduced flying activity and olfactory discrimination, and it also affected olfactory learning skills. It has been demonstrated that imidacloprid exposure at field levels decreases colony development and new queen production. According to a 2012 in situ study, honey bees fed high fructose corn syrup (HFCS) in situations where forage is scarce were exposed to sublethal amounts of imidacloprid, which resulted in symptoms resembling CCD in the bees. The HFCS-fed control group that did not receive pesticide does not exhibit these symptoms. Neonicotinoids were detrimental to bees in the usual dosages, and substitutes were desperately sought. The quantities that the bees ingested were not fatal, but prolonged exposure might result in long-term issues.

Field Studies

Scientists discovered in 2005 that pollen from seeds treated with imidacloprid had significant quantities of the insecticide. They speculated that polluted pollen could be the reason of honey bee colonies dying. Imidacloprid has sublethal effects on honey bees, according to laboratory research, however there is insufficient data to support the findings' relevance to the environment, according to a meta-analysis study released in February 2010. Similar to this, a 2012 study found that while honey bees treated to neonicotinoids in a lab setting showed both deadly and sub-lethal effects on their foraging behavior, memory, and learning capacity, similar effects were not observed in field experiments using dosages that were representative of the real world.

Numerous studies revealing previously unrecognized pathways of neonicotinoids exposure impacting bees—such as dust, pollen, and nectar—were published in 2012. Studies have also demonstrated environmental persistence in soil and irrigation channels used in agriculture. Another potential exposure source is the air, which is released by the machineries used to sow maize seeds coated with imidacloprid and clothianidin.An 11-year British study published in 2015 found a link between rising neonicotinoid seed treatment use in agriculture and rising losses of honey bee colonies on a landscape scale. The association between neonicotinoids and CCD was originally demonstrated in this field investigation.

Regulatory Policy

The relationship between imidacloprid and the "dying bees" phenomena has been the subject of discussion for a while in Europe. A partial ban on imidacloprid was implemented in France as a result of a discussion centered around a research from the "Comité Scientifique et Technique (CST)". The French Minister of Agriculture, Jean Glavany, outlawed the imidacloprid pesticide Gaucho in 1999, mainly because of worries about possible harm to honey bees. While French beekeepers were successful in getting neonicotinoids outlawed, the Clinton administration approved imidacloprid and other previously outlawed pesticides. In 2004, the Bush administration increased the use of pesticides while further reducing controls.

A formal study conducted in 2013 by the European Food Safety Authority (EFSA) revealed many data gaps that had not been taken into account before and concluded that some neonicotinoids posed an unacceptable high risk to bees. They came to the following conclusion after reviewing the data: "Dust drift exposure to seed treatments used in maize, oilseed rape, and cereals was found to pose a high acute risk to honey bees." Additionally, exposure via residues in nectar and/ or pollen was found to provide a high acute risk."

The European Commission (EC) suggested a two-year ban on neonicotinoids following the 2013 EFSA review. The European Union decided to prohibit the use of neonicotinoid pesticides for two years in April 2013. The use of clothianidin, thiamethoxam, and imidacloprid on crops that attract bees was restricted by the prohibition. The British government was one of the eight countries who voted against the motion, claiming that the research was insufficient. As stated at the 1992 Rio Conference on the Environment and Development, the "precautionary principle" states that "lack of full scientific certainty shall not be used as a reason for postponing cost-effective measures to prevent environmental degradation." This ban is an implementation of this principle.

The United States Environmental Protection Agency (EPA) was sued in 2013 by beekeepers and environmentalists for allegedly permitting the use of neonicotinoids in the country. The suspension of thiamethoxam and clothianidin was particularly requested in the lawsuit. In response to the lawsuit, the EPA cited studies that demonstrated the exaggerated impact of neonicotinoids in CCD and implicated the Varroa mite in the reduction in bee populations. In Congress, the Save America's Pollinators Act (H.R. 2692) was first presented in 2013 and then again in 2015. Neonicotinoids were to be put on hold until their effects had been thoroughly reviewed, according to the proposed statute.

Fipronil

Fipronil is a broad-spectrum systemic insecticide of to the phenylpyrazole class. Fipronil is intended to eradicate bee-like insects, including yellowjackets (Vespula *germanica)* and numerous other colonial pests, by a method known as "toxic baiting," in which a single insect that returns to the hive disperses the pesticide among its offspring.

The French Directorate-General of Food at the Ministry of Agriculture concluded in May 2003 that acute fipronil toxicity was responsible for a mass bee mortality case that had been reported in southern France. Dust was produced during the improper seed treatment, which was connected to toxicity. The ministry made the decision in February 2003 to temporarily halt the sale of BASF crop protection products—including the "Regent" brand—that contained fipronil in France.

The Minutes-Association for Technical Coordination Fund in France conducted a study in 2010 that linked fipronil to the spread of CCD in bees. The study found that even at very low nonlethal doses, this pesticide made it more difficult for foragers to locate the hive, which led to a significant loss of foragers on each pollen-finding expedition. However, no mention was made of any other symptoms associated with CCD. Subsequent research has shown that fipronil has no acute effects on honey bees.

Fungicides

In 2013, scientists fed healthy bees pollen that they had collected from colonies. On average, nine distinct fungicides and insecticides were present in the pollen.

The researchers also found that there was a threefold increase in the likelihood of parasite infection in bees that consumed pollen contaminated with fungicides. According to their research, fungicides—which are generally believed to be safe for bees—may potentially be a major factor in CCD. The bees the scientists collected almost entirely foraged from weeds and wildflowers, not from crops, indicating that bees are more commonly exposed to pesticides than previously believed. This suggests that spraying techniques may need to be reassessed. As stated by University of Maryland ecologist Dennis van Engelsdorp, "Fungicides, which we didn't expect to harm insects, seem to have a sub-lethal effect on bee health." He continued by saying that this is significant because fungicides are not subject to strict regulations.

Antibiotics and Miticides

The majority of beekeepers afflicted with CCD state that they employ antibiotics and miticides in their colonies; nevertheless, it is doubtful that any one of these chemicals is the cause of the problem due to the variation in the specific chemicals used. But it's possible that some of these commonly used pesticides haven't been examined for potential impacts on honey bees, and as a result, they might be exacerbating the CCD situation.

Antibiotics

According to a University of Texas in Austin study, antibiotics that are frequently used in beekeeping to treat disease have been shown to diminish the gut bacteria levels in honeybees, increasing their susceptibility to illness. Colony collapse can occur when viral infections spread widely within a beehive. Researchers say that even though the antibiotics are essential, beekeepers should use prudence in deciding how much and how long to administer them. Colony collapse can occur when viral infections spread widely within a beehive.

Miticides

Miticides are used by beekeepers to eradicate Varroa infestations from their colonies; nevertheless, therapy may result in increased viral infections in the colonies. Honey bees may experience immunological suppression as a result of high treatment dosages or prolonged miticide use, which increases their susceptibility to viruses.

In 2008, Pennsylvania State University researchers discovered that samples of wax from hives included significant concentrations of the miticides fluvalinate and coumaphos, along with lower concentrations of seventy other pesticides. Decreased survival rates of developing queens have been linked to elevated concentrations of the organophosphate Coumaphos in wax. Under the direction of entomologist Steve Sheppard, researchers from Washington State University verified high concentrations of pesticide residue in hive wax in 2009 and discovered a link between it and noticeably shorter bee lifespans. The concentrations of chemicals in wax and pollen were also higher in a major 2010 survey of healthy and CCD-affected colonies, but they were comparable in failing and healthy hives.

Climate Change

Although it is unknown how specific prospective environmental changes brought on by climate change may affect honey bee growth, they may have an impact.

Bee Rentals and Migratory Beekeeping

Ever since American beekeeper Nephi Miller started relocating his hives during the winter of 1908, migratory beekeeping has expanded throughout the United States. American agriculture would not be able to produce at all at its current levels with native pollinators alone, making bee rental for pollination an essential component of the industry. The total revenue earned by American beekeepers from renting their hives for pollination is significantly more than that of honey production.

Researchers worry that transporting bee colonies across the nation to pollinate crops, where they interact with other bee colonies from other regions, may contribute to the transmission of viruses and mites between colonies. Furthermore, some believe that this constant relocation and migration strains and disturbs the hive as a whole, possibly making it less resilient to various forms of systemic illness. Not only are viruses and mites a problem when moving bees across the nation, but stress during transportation may also be a contributing factor in colony collapse disorder.

Selective Commercial Breeding and Lost Genetic Diversity in Industrial Apiculture

Environmental factors have received the most attention in the context of CCD. 'Industrial' or agricultural areas where commercially bred bee colonies are used are known to be the areas where CCD has the worst impact. Large within-colony populations of bees result from the complicated and highly selective process of natural breeding and colony reproduction in wild bees. This genetic diversity may not be replicated in commercially developed colonies.

Malnutrition

The CCD Study Group at Pennsylvania State revealed in 2007 that a time of "extraordinary stress" impacting the colonies in question before to their die-off was recognized by all producers in a preliminary survey. This stress was typically related to inadequate nutrition and/or drought. Since this was the only thing that the report's examples of CCD shared, there seemed to be a good chance that the phenomena was tied to nutritional stress, which might not be evident in colonies that are healthy and fed.

According to a 2010 study, bees given pollen from a range of plant species had immune system healthier indicators than bees fed pollen from a single species. Even though the pollen had a higher protein content, bees fed pollen from five different species had higher levels of glucose oxidase than bees fed pollen from only one species. The authors postulated a potential connection between CCD and a decline in plant diversity. A healthy population of honey bees can be achieved by a good diet, according to research. "The authors recommended a diet containing 1000 ppm potassium, 500 ppm calcium, 300 ppm magnesium and 50 ppm each of sodium, zinc, manganese, iron and copper."

According to a 2013 study, p-coumaric acid, which is often contained in honey, helps bees detoxify certain pesticides. Therefore, its lack from artificial nutrition given to bees may be a factor in CCD. Subsequently, a 2014 study discovered that, in contrast to bees given comparable low-protein honey, those fed high-fructose corn syrup or sugar displayed down regulation in numerous genes linked to protein metabolism and oxidation reduction.

Electromagnetic Radiation

There have been a lot of discussions on the Internet and in the community at large, but very few thorough research on the effects of electromagnetic field exposure on honeybees have been published in peer-reviewed scientific literature. In 1981, one of the few peer-reviewed studies was published, revealing that bees were not seriously affected by microwave radiation, even at far higher strengths than those employed in communication. findings of a study on the non-thermal effects of radio frequency (RF) on honey bees (*Apis mellifera* carnica), exposure to DECT cordless phone base stations operating at 1,880–1,900 MHz did not cause any changes in behavior in the bees. A further study found that bees may be less able to return to their hive when exposed to close-range electromagnetic fields (EMF).

Half of their colonies collapsed during the trial, including some control hives without implanted DECT base stations. News of this study first surfaced in April 2007 in a number of media sites, starting with a report in The Independent that claimed the study's subject matter covered mobile phones and had connected them to CCD. Cell phones were not included in the cited study, despite the fact that they were at the time linked to other media reports. Since then, the researchers involved have clarified that neither their research nor its findings relate to CCD. They also claimed that the Independent article misconstrued their findings and turned them into "a horror story".

Experts evaluated 919 peer-reviewed scientific papers examining the effects of electromagnetic fields (EMF) on humans, animals, and plants. Of these, 6 reported adverse consequences from EMF radiation exposure, but none specifically linked to chronic cluster headache disorder (CCD). An investigation of the non-thermal impacts of electromagnetic exposure on learning was carried out in 2004. The DECT base station operates between 1880 and 1900 MHz, but the investigators did not observe any change in behavior as a result of RF exposure. Low-frequency magnetic fields, or weak static, are detectable by honeybees and are one of numerous cues they employ for navigation. Apart from slight heating effects, no mechanism has been found to influence the behavior of insects by weak radio frequency energy.

Genetically Modified Crops

It is not widely accepted that GM crops cause CCD. Honeybee larval and adult survival was not adversely affected by Bt proteins used in commercialized genetically engineered crops to control lepidopteran and coleopteran pests, according to a 2008 meta-analysis of 25 independent studies evaluating the effects of Bt Cry proteins on honeybee survival (mortality). Furthermore, just a small percentage of the larvae's protein was obtained from pollen, and there was no geographic association between the sites of GM crops and the places where CCD was observed.

Management

While beekeepers noticed indications of CCD, the Mid-Atlantic Apiculture Research and Extension Consortium (MAAREC) tentatively recommended the following actions as of March 1, 2007:

1. Don't mix strong colonies with failing colonies.
2. If you discover a collapsed colony, put the equipment in a location where you can take precautions to keep bees from accessing it.
3. Use fumagillin if you feed sugar syrup to your bees.
4. Treat the colonies with oxytetracycline rather than tylosin if you observe a secondary illness, such as European Foulbrood, and colony collapse.

Simply using native bees like bumble bees and mason bees instead of hiring beekeepers is another suggested solution for farmers of pollinated crops. By offering appropriate nesting sites and some extra crops the bees could use for food (for example, after the commercial crops on the farm's pollination season has concluded), one can aid native bees in establishing themselves. An effective varroa mite-resistant variety of bees was created by a British beekeeper. Russian honey bees have unfavorable characteristics that reduce their usefulness in commercial apiculture, even if they also resist varroa mite infestations. Nevertheless, they are still vulnerable to other variables linked to colony collapse disorder.

A 15% decline in the bee population over the preceding two years led to the creation of a national bee database in the United Kingdom in March 2009 to track colony collapse. The Department for Environment, Food, and Rural Affairs is funding the register, which is run by the National Bee Unit. Its main purposes are to track health trends and determine whether colony collapse disorder poses a threat to the honey sector. The 20,000 beekeepers in Britain have been invited to take part. "We still do not believe CCD is a cause of colony losses in the UK, however we are continuing to experience colony losses, many if not most of which can be explained," said David Aston, the British Beekeepers' Association, in October 2010. Raising the profile of integrated bee health management—that is, identifying and attempting to eliminate variables that lower a colony's health status—is the strategy being used in UK beekeeping.

Economic and Ecological Impact

Since honey bees are not native to the Americas, their use as pollinators in the United States and other Western Hemisphere regions is restricted to strictly agricultural and ornamental purposes. No native plant requires honey bee pollination, with the exception of situations where honey bee populations are concentrated in monocultures, where the need for pollination during blooming is so great that native bee populations cannot support the concentration of pollinators (with current technology).

This is especially significant for crops like almonds grown in California, where honey bees are the primary pollinators and the harvest was valued at $3.6 billion in 2011. It was projected in 2000 that the value of all U.S. crops that relied only on honey bee pollination exceeded $15 billion. The production of almonds in California grew by more than 30% in just the last ten years, from 370 million pounds in 1995 to a record 2,500 million pounds in 2019.The almond business in California rents about 1.6 million honey bee colonies in the spring to pollinate their crop, a result of the strong demand for pollinators driving up the cost of renting honey bees. Honeybees provide pollination services worth about $200 billion to the world economy.

About one-third of the crop species grown in the United States, including watermelons, cantaloupes, cucumbers, strawberries, raspberries, blackberries, cranberries, almonds, peaches, apples, pears, cherries, and cucumbers, are pollinated by them. In the United States, other insects, such as bees (like squash bees on cucurbits), can and frequently do pollinate many, but not all, of these plants; however, this is usually not done on a commercial basis. Although some farmers use honey bees to aid in pollination for a few types of local crops, none of them are very dependent on them, and when honey bees disappear from an area.

Given that native pollinators are usually better suited to support those plants, there is a chance that they will retake the niche (given that the plants regularly occur in that specific location). On the 30% of crop types where honey bees are employed, most native pollinators cannot be mass-utilized as easily or as effectively as honey bees—in many cases, they will not visit the plants at all—despite the fact that many other species are actually more efficient at pollinating on an individual basis. Beehives are mobile and can be moved from crop to crop as needed. Saturation pollination is how bees make up for their inefficiencies by visiting a huge number of plants. This means that the beekeeping sector is closely linked to the commercial viability of these crops. Apple orchard hand pollination in China is an expensive, time-consuming, and labor-intensive process.

Having said that, honeybees pollinate over 75% of all plant species that are directly used for human consumption globally. It is predicted that seven of the 60 major agricultural crops that contribute significantly to the North American economy would disappear in the event of a catastrophic loss of honeybees, and this is only for one area of the planet. Due to their need on honeybees, farms with intensive systems—high crop densities—will be most affected, as opposed to non-intensive systems, such as little backyard gardens that rely on wild bees. The pollination services provided by honeybees, which alone cost $1.25 billion in the United States each year, are in high demand from these kinds of farms. However, this expense is covered by the 22.8 to 57 billion euros that honeybees produce worldwide as pollinators.

CHAPTER 13
Honey Quality Testing

Bees gather honey from two distinct sources: honeydew and nectar. The most widely used form of honey worldwide is nectar, whereas honeydew is mostly found in European nations like Greece and Austria. A sugar solution with variable concentrations released by flower nectaries is called nectar. The main sugars in nectar are sucrose, glucose, and fructose; this sugar composition is characteristic for the plant species. The primary sugars in nectar are fructose and glucose. In reality, a variety of climatic conditions, including temperature, soil type, humidity, and season, affect the sugar content. Honey is a syrup high in carbohydrates. It has been utilized as a folk remedy from ancient times in addition to being a well-liked sweetener.

The phenomenon of adulterating honey and producing synthetic honey is a result of the market's increased need for honey these days. The price of synthetic or contaminated honey is significantly less than that of pure honey, which is due to the seller's intense desire for profit. The act of incorporating some foreign materials into pure honey is referred to as "honey adulteration." There had been this incidence for hundreds of years. The following was mentioned in Sir John Hill's 1957 first book, "The Virtues of Honey," which warned readers: "Beware of honey with dishonest mixture or flour and other ingredients." It had become a common problem in the market nowadays due to the difficulties in identifying the adulterated honey.

Throughout the years, a variety of foreign materials have been utilized as adulterants in honey adulteration, including wheat, paraffin, glucose, dextrose, molasses, corn syrup, and sugar syrup. Products containing starch and comparable substances may cause some physical characteristics to change, such as samples that taste entirely sweet, have an odd smell, or are incredibly light or dark in color. Adulterants could be easily identified in the early 1970s before high fructose corn syrups were developed. A substance with properties akin to those of pure honey is called high fructose corn syrup (HFCS). It is frequently offered at a lower cost than pure honey and includes the majority of the main ingredients. Since then, the possibility of adding 80 to 90% of HFCS without being detected has made honey adulteration with HFCS exceedingly profitable.

Codex Alimentarius Standard for Honey

A comprehension of the numerous quality criteria is provided in this study. The following quality criteria are determined using the standards and procedures included in the standard drafts: water-insoluble matter, hydroxyl methyl furfural, apparent reducing sugars, apparent sucrose, moisture, ash, acidity, and diastase activity. But in the past 30 years, there haven't been many published studies on lowering honey's sugar and ash levels. Rather, the main ingredients are electrical conductivity and certain sugars. International honey standards for the total amount of fructose and glucose, sucrose content, and electrical conductivity are suggested based on this facts. In addition, the application of other quality factors—also employed in many nations—such as proline, invertase activity, and particular rotation is covered.

Moisture Content

The sole compositional requirement that must be met in order for the global honey trade to comply with the Honey Standard is moisture content. High water content honey has a higher propensity to ferment. The draft for a new standard suggests a maximum value of 21 g/100g. The measurements from recent years do not support the exception for clover honey. As a result, clover honey's maximum water content should likewise be 21 grams per 100 grams. Values as high as 21 g/100 g are incredibly uncommon in real-world situations. Ninety-one to ninety-five percent of all honeys had a water content of less than 20 g/100 g in regular honey control conducted by the IHA on almost 30,000 honey samples between 1989 and 1997 (8) A 20 g/100 g criterion was also effectively used in Switzerland for the previous 20 years, up until the most recent modification of the Swiss Food Ordinance, which required acceptance of the maximum value of 21 g/100 g set by the European Union. A lot of national beekeeping organizations (such those in Germany, Belgium, Austria, Italy, Switzerland, Spain, or other countries?) have set maximum moisture content levels for certain quality honey classes, ranging from 17.5 to 18.5 g/100 g.

Apparent Sugar Content

The vast majority of honey sugars in most flower honeys are represented by apparent reducing sugars; however, this is frequently not the case in honeydew honeys. In fact, non-reducing oligosaccharides including melezitose, maltotriose, and raffinose are present in high concentrations in a lot of honeydew honeys. These results led to a modification of the apparent sugar standard in the Codex draft with respect to the prior standard: a minimum of 45 g/100 g has been proposed, as opposed to the previous standard's minimum of 60 g/100 g. The former standard of 60 g/100 g is maintained in the European draft.

True sucrose and "apparent sucrose" are frequently not the same. "Appearant sucrose" is assessed indirectly by comparing total and reducing sugars. In this case, the two drafts are comparable; the Codex document contains more varieties of honey than the European one. The IHC suggests making an exemption in light of rosemary honey's high sugar content. In fact, tests on 33 rosemary honeys from Spain revealed that a sizable percentage of these honeys had more than 5% apparent sucrose. Measuring reducing sugars only distinguishes between blossom and honeydew honeys; however, other techniques, such as measuring electrical conductivity, make this distinction much easier to identify. There are numerous arguments in favor of using the measurement of specific sugars in place of the reducing sugars.

Water Insoluble Solids Content

Detecting insoluble matter is a crucial tool for finding pollutants in honey that exceeds permitted limits. It took place in an era when pressing combs produced a large amount of the world's honey. However, centrifugation is now used to extract practically all commercial honey. We believe that the 0.1 g/100 g allowed maximum under the Codex and European standards is excessively high. The majority of the readings are lower, falling between 0.005 and 0.05 g/100 g. One of the main sources of water-insoluble contamination is wax, which is not determined by the Codex approach. Other filtration techniques, such as paper filters, can be employed for this purpose, however they haven't been formally presented yet.

Ash Content

The botanical origin of honey is determined by its ash content, with blossom honeys having a lower ash level than honeydew honeys. These days, electrical conductivity measurements typically take the place of this one. While conductivity is being recognized as a global standard, the ash content may continue to be considered quality criteria.

Acidity

Acidity is a crucial criterion for quality. A maximum acidity value has been found to be helpful due to the increased acidity caused by honey fermentation, despite the significant natural fluctuation. Since certain honeys have a higher natural acidity than others, the former criterion, which set a maximum of 40 milliequivalents/kg, has been raised to 50 milliequivalents/kg in the Codex draft.

Diastase Activity

As a quality component that is impacted by both heating and storage, honey diastase activity serves as a gauge for both overheating and freshness. Even though diastase naturally varies widely, the current standard of a minimum DN value of 8 has shown to be helpful. Over 92% of the raw honey samples (n = approximately 20,000) and over 88% of the retail honey samples (n = approximately 1000) in the long-term routine honey control at the IHA had a DN greater than 8 (8). It is important to realize that some unifloral honeys naturally have low diastatic activity when evaluating diastase results. The EU standard is applicable to the entirety of retail honey, whereas the Codex standard only relates to honey that has been processed and blended. Despite this, the minimal requirements for diastase activity in the EU drafts and the Codex are identical in practice. Since it is anticipated that diastase activity will decrease with storage, the EU standard is actually more stringent than the Codex rule.

Hydroxy Methyl Furfural (HMF) Content

This important component of honey quality indicates both overheating and freshness. Hydroxymethylfurfural (HMF) is almost nonexistent in fresh honeys, but it can rise during storage based on the honey's pH and storage temperature. Certain European bee federations (Belgium, Germany, Italy, Austria, Spain) sell up to 15 mg/kg of "quality honey" from their honey. A maximum value of 40 mg/kg has shown to be satisfactory in international trade.

Over 90% of the raw honey samples (n = 30,000) and over 85% of the retail honey samples (n = 2000) in the long-term routine honey control at the IHA over the last ten years had less than 30 mg HMF/kg (8). The proposed maximum for Codex is 60 mg/kg. The idea of raising the maximum value is founded on the observation that honey storage in nations with warm climates results in higher HMF values. A limit of 40 mg/kg is required under the most recent EU standard proposal, as this standard has been shown to be effective in European settings. The two rules differ significantly in this regard as well, just as they do with diastase: the EU standard covers all retail honey, but the Codex standard only applies to honey that has been processed and blended. Given that HMF is predicted to rise during storage, the EU standard is actually far harsher than the Codex guideline.

Electrical Conductivity

Conductivity, rather than ash content, is now used in routine honey control as a good indicator of the honey's botanical origin. The ash and acid content of the honey determines this measurement; the larger their amount, the higher the conductivity that results. The electrical conductivity and ash content have a linear relationship:

$$C = 0.14 + 1.74\ A$$

where C is the electrical conductivity in milli Siemens cm-1 and A the ash content in g/100 g.

Recently, extensive conductivity data covering thousands of commercial honey varieties were released. We suggest that honeydew and chestnut honeys should have more than 0.8 mS/cm, and blossom honeys and combinations of blossom and honeydew honeys should have less than 0.8 mS/cm based on these findings. The honeys from Arbutus, Banksia, Erica, Leptospermum, Melaleuca, Eucalyptus, and Tilia, as well as their blends, are an exception because of their incredibly wide conductivity variations.

The conductivity measurement is quick and simple, requiring only low-cost equipment. It is often utilized to distinguish between honeydew and blossom honeys as well as to describe unifloral honeys. Therefore, it is urgently advised that a worldwide conductivity standard be introduced.

Specific Sugar Content

A broad guideline for the minimum fructose and glucose content of 60 g/100 g for all blossom honeys and 45 g/100 g for all honeydew honeys can be suggested based on the substantial data that we just published. In over 99% of the honeys that were analyzed, this requirement could be met. The case of sucrose is more intricate. In this case, over 99% of the honeys that were analyzed met the general threshold of 5 g/100 g. As fructose and glucose together account for more than 90% of all reducing sugars, the total amount of these two sugars is extremely near to the total amount of all sugars. In fact, the suggested minimum limits for apparent reducing sugars of 45 and 65 g/100 g, respectively, and the sum of glucose and fructose of 45 and 60 g/100 g for honeydew and blossom honeys are nearly comparable. However, the suggested standard for genuine sucrose and the one for apparent sucrose are extremely similar.

A few Australian and New Zealand honeys are included in the standard for reducing sugars but not in the proposed standard for specific sugars because no specific sugar data is available for them. Another exception is for honeydew honeys, where the "apparent sucrose standard" is 15 g/100 g while the specific sucrose standards are only 5 g/100 g.

Other benefits for routine honey management will result from the establishment of a standard for particular sugar amounts. Thus, two useful parameters for distinguishing between various unifloral honeys are the fructose/glucose ratio and the sucrose concentrations. Additionally, the concentration of certain higher sugars such as melezitose and maltotriose can serve as reliable indicators of the honeydew content. Information about sugar adulteration and genuine honey can also be obtained from the particular sugar spectrum.

Invertase Activity

As a freshness indicator, invertase activity is extremely sensitive to heat and storage damage. An invertase number (IN) of greater than 10 was suggested for fresh, unheated honeys; an IN of greater than 4 is advised for honeys with low enzymatic activity. While invertase activity exhibits significant natural variation, similar to honey diastase, its utility in honey quality management has been demonstrated. The honey standards of the German, Belgian, and Spanish beekeepers associations also use freshness invertase criteria.

Proline Content

A measure of the ripeness of honey and, in certain situations, the presence of adulterated sugar is the honey proline content. In honey control laboratories, 180 mg/kg is the minimal figure that is considered to be authentic honey. But keep in mind that depending on the type of honey, there is a significant variance in proline.

Specific Rotation

The values of the various honey sugars combine to provide the total value for the optical rotation. Currently, Greece, Italy, and the UK employ particular rotation measures to differentiate between blossom and honeydew honeys. It was discovered in Italy that although honeydew honeys have a positive optical rotation value, blossom honeys have a negative one. Future research has to determine whether this technique can distinguish between various honeys in different geographical areas.

Techniques of Honey Purity Tests

1. Determination of HMF level by spectrophotometer. 2. Determination of sugar profile by gas chromatography 3. Determination of water content by hand held refractrometer. 4. Determination of hydrogen peroxide. 5. Determination of pH value by pH meter.

If the honey samples satisfy every requirement for being considered pure honey, they are classified as such. Generally speaking, honey originating from tropical regions like India has a minimum of 60% reducing sugar (fructose, glucose, and maltose).Fructose to glucose ratios vary from 0.9 to 1.35. Usually, less than 5% of it is sucrose. Tropical honey has a little greater HMF content, but it still doesn't exceed 80 mg/kg. Although the adulterated honey sample's purity test results were comparable to those of the pure honey, there was more than 5% of sucrose present. In contrast, the synthetic honey sample has a high concentration of HMF—more than 80 mg/kg.

Determination of Hydroxymethylfurfural (HMF) Level

The concentration of 5-(hydroxymethyl-) furan-2-carbaldehyde (HMF) was ascertained using this technique. The HMF level result is commonly given in milligrams per kilogram (mg/kg). To assess the HMF content, 5.0g of each honey sample was precisely weighed into a beaker. After that, 25.0 ml of distilled water was added, and everything was thoroughly mixed until the 5.0 g of honey samples had fully dissolved. After that, the combined solution was put into a 50 ml volumetric flask. The volumetric flask was filled with 0.5 ml of Carrez solution I, and it was well mixed by vortexing. After adding Carrez solution II to the combined solution, it was thoroughly mixed using a vortex. After that, the volumetric flask was filled to the brim with distilled water. It may be necessary to add a drop of ethanol to reduce the froth that forms when mixing. After then, filter paper was used to filter this combination. After filtering, the first 10 milliliters of the solution were discarded, and the remaining solution was saved. The filtrated solution was pipetted into two different test tubes, each holding 1.0 ml of volume.

The test tube containing the honey solution was filled with 1.0 milliliter of distilled water for the sample solution. After thoroughly mixing the solution with a vortex, 1.0 milliliter of 0.2% sodium bisulphate was added and thoroughly mixed as a reference. Using a 1cm quartz cell, the absorbance of the sample solution at 284 nm and 336 nm was measured against the reference solution in less than an hour. In order to get a sample absorbance low enough for accurate results, distilled water and 0.2% bisulphite were added to the reference if the absorbance at 284 nm was more than the value of 0.6. Every sample of honey was prepared and put through three tests. S.E.M. (mean plus standard error mean) was used to record the results.

Determination of Sugar Profile

The sugar composition of the honey samples (glucose, fructose, sucrose, and maltose) was ascertained using this approach. 1.0 mg of each standard sugar (fructose, glucose, maltose, and sucrose) was precisely weighed into a 5 ml screw-top container to prepare the sample for standardization. HMF concentration (mg/kg) is equal to absorbance at 284 nm (A284-A336 x 149.7 X 5 x D W A284). The absorbance at 336 nm is A336. D = Dilution Factor 149.7 = Factor 126x1000x1000 16830x10x5 W = Sample weight 126 is the HMF molecular weight. Molar absorpticity of HMF at 284 nm = 16830. G conversion to mg = 1000. 10 = 5 Converts into 50 milliliters. 1000 = One kilogram of honey divided by thirty-six glass vials.

A 5 ml screw-top glass vial containing 5.0 mg of precisely weighed honey samples was used for the fabrication of the honey sample. After that, 0.45 milliliters of pyridine were added to the produced samples. For a minimum of ten minutes, these vials were submerged in the waterbath at 70 C. Each vial was filled with 0.5 ml of hexamethyldisilazane (HMDS) and thoroughly mixed. Subsequently, 0.05ml of Triflouroacetic acid (TFA) was cautiously added to the mixture, drop by drop. Periodically, the vial was shook while the TFA was added at room temperature.

We let these transparent solutions stand for a minimum of fifteen to thirty minutes. It produced a clear, homogenous solution. It was necessary to prepare fresh samples in case the solutions get hazy. Prior to injecting the solutions into the gas chromatography (GC-14A), they were allowed to stand at room temperature for a whole day. Each sample solution has to be injected into the gas chromatography column's capillary in 1.0µl increments.

Calculation for weight and percent of sugar in honey.

$$\text{Weight of sugar in honey (mg)} = \frac{\text{Peak of sugar in Honey}}{\text{Area Peak in Standard}} \times \text{Weight of standard } \mathbf{(mg)}$$

$$\text{Per cent of Sugar in honey (\%)} = \frac{\text{mg sugar in honey}}{\text{Weight of honey sample}} \times 100\%$$

Determination of Water Content in Honey

The purpose of the handheld honey refractometer is to ascertain the proportion of water present in honey . The process of determining the water content of honey was simple when utilizing this honey refractometer. Before being utilized, the hand-held refractometer's prism was cleaned and dried. Subsequently, a honey sample was uniformly applied to the refractomer's prism . The refractive index was measured and noted. After measuring each sample twice, the average value was determined. The refractometer's prism was meticulously cleaned after usage.

Determination of Hydrogen Peroxide

This test, which finds out whether hydrogen peroxide is present in the honey sample, is the simplest to run. After weighing 3 grams of honey and diluting it with 10 milliliters of distilled water, a 30% (w/v) concentration of honey was created. After that, the mixture was incubated for 30 minutes at 37°C in a waterbath 38. To determine the concentrations of the H_2O_2 produced, test strips from the test kit were dipped in the mixture and the color developed was compared to the color code. For the sake of accuracy, each sample was tested three times, and the findings were noted.

Determination of pH Level

The crude honey's pH level was incredibly erratic and difficult to gauge. Consequently, honey was diluted with a 10% (w/v) solution. Prior to usage, the pH meter was calibrated at pH 3.7 (4.0), 7.0, and 9.0. A sample of 10 g honey was dissolved in 75 ml of distilled water without carbon dioxide. A magnetic stirrer was used to combine this solution. The solution was submerged beneath the pH electrodes. The honey solution's pH was measured to two decimal places.

Honey Purity Test

However, a typical consumer may find it easier to determine if the honey they are purchasing is real or phony with the use of a few simple guidelines. These pointers are based on the qualities listed in detail below. Still, it takes some practice to identify contaminants in honey.

TEST	PURE HONEY	FAKE HONEY
Stickiness	Not sticky if rubbed between fingers	Sticky due to additional sweeteners
Thickness	Fairly thick, takes time to move from one side of the jar to the other	Very light and runny
Taste	Taste vanishes in a matter of minutes (heating and cooling many times may alter the taste)	Taste remains for a little longer because of the presence of sugar
Smell/Aroma	Mild scent, probably the actual smell of the flowers from which the nectar was collected	None or sour smell
Heating	Caramelizes quickly and doesn't not form foam	Becomes bubbly and never caramelizes
Dissolving Method	Doesn't get dissolved in water, will lump and settle at the bottom. Gets diluted when stirred for a while. Mixing in equal amounts of honey and methylated spirits, honey settles at the bottom.	Stays incoherent and gets dissolved water right away. Dissolves in methylated spirits while making the solution milky.

Flame Test	Matchstick lights even and easily after it has been immersed in pure honey	Matchstick doesn't light easily due to presence of moisture
Bread Test	When spread on a slice of bread, the slice hardens in minutes	Wets the bread due to moisture content
Absorption Test	Few drops poured on bloating paper doesn't get absorbed. When poured on piece of white cloth, it doesn't leave stains	Gets absorbed in bloating paper. Leaves stains on a white piece of cloth
Impurities	Presence of impurities: dirty-looking particles, pollen and bee body parties	Absence of impurities

Most people consider honey as a preservative. However, beekeepers frequently overlook or forget that honey is a perishable product in addition to being a sensitive natural good. The quality of honey can quickly decline with improper handling. Honey is a product used as food and medicine, thus every conscientious beekeeper, honey processor, and honey packer must always remember that it requires hygienic care from the time it is extracted until it is packaged and supplied to the consumer.

Therefore, treating this natural commodity at all levels requires more attention and care due to the honey industry. This is no longer a question of personal taste; rather, it is now required by law due to the Food Adulteration Act's severe penalties and strict enforcement across the globe. The most crucial element is the moisture content of the honey since it influences its ability to be kept fresh or encourages fermentation. The honey is uniquely ripened and sealed by the bees. Beekeepers should ensure that he only extracts honey that is matured and sealed. On average, 40% of the nectar is made up of sugar, with the remaining 60% being water. Nectar is supplied to the hive bees upon the forager's return to the colony, whereupon the conversion of sucrose proceeds. When the forager gives the honey to the hive bees, some of the water in the nectar is also taken out. Furthermore, when honey is within the cells, it ripens. The honey's ability to mature is contingent upon the level of cell filling. If more than half of the cells are packed with unripe honey, further time is needed. Because of this, it is ideal for the honey chamber to hold all of the super frames, or at most one less than its maximum capacity.

Sliding super slightly forward to enable cross ventilation can help provide additional ventilation in dry areas during the rapid honey flow season. This will hasten the honey's ripening process. The location's relative humidity has an impact on how quickly food ripens. The relative humidity in many honey-producing regions of India, especially the Southern and Eastern States, is frequently higher than 60%, and the daytime room temperature is nearly identical to the hive temperature (34 C). In these circumstances, ripening proceeds quite slowly.

Granulation in Honey

Although most honeys eventually granulate, freshly extracted honey is in liquid state. The primary determinants of granulation are the nectar source and the handling of the honey during processing, packing, etc. The granulation might be partial or uniform, and it can take many different forms, from extremely smooth to coarse crystal. Honey, when it comes to the dextrose sugar it contains, is a supersaturated solution. Put another way, this indicates that there is more dissolved dextrose than would typically be possible in the water found in honey. As a result, the solution is unstable and tends to stabilize when some glucose (dextrose) crystals develop.

Before any crystal can form in super-saturated fluids, crystallization must first begin on a base, or nucleus. More molecules are drawn to the crystallization process once it begins, and eventually a discernible form is seen. The honey will granulate more quickly and uniformly the more nuclei it contains. Nuclei for the granulation process are provided by air bubbles, foreign particles, and an abundance of pollen grains. For optimal consumer appeal, the honey has to be completely granulated or liquid. Granulated honey is occasionally seen as tainted honey in India since liquid honey is more in demand than granulated honey.

When processing honey with a high glucose percentage, a slight increase in temperature (65 to 68° C) might be used to postpone granulation. The honey must then be completely filtered.

Fermentation

Under specific circumstances, all biological products experience chemical transformations that result in the breakdown of molecules into simpler compounds. Most people refer to this process as fermentation. Since honey is a biological product, improper handling can cause it to ferment or spoil. Honey ferments because of the presence of fermenting yeasts, which are found in practically all honeys. Honeys with high sugar concentrations or low moisture content keep yeasts latent or inactive. When honey is unripe and has a moisture level of more than 22%, the yeasts begin to grow and ferment the honey. Due to partial granulation, even in ripe honey, the liquid portion's moisture level frequently exceeds 25%, which triggers fermentation. The major reasons honey ferments are high moisture content or incomplete granulations.

Processing of Honey

Honey is heated to stop fermentation and delay granulation. It is advised to heat honey between 140 and 145 degrees Fahrenheit (60 and 62 degrees Celsius) for a few minutes, without sacrificing its flavor or color or causing its constituents to break down, because this is the temperature at which the yeast cells that ferment honey are eliminated. However, during processing, the color and flavor of honey are affected by the presence of foreign materials, particularly beeswax.

This is due to the fact that heating honey to a temperature that coincides with beeswax's melting point causes coloring and flavoring ingredients from melted beeswax to permeate into the honey and degrade its quality. Thus, straining should be done on all extracted honeys prior to processing. However, honey's high viscosity makes it difficult to strain quickly. As the temperature rises, the viscosity of the honey rapidly drops. Therefore, it is advised to warm honey to a temperature of 100 to 110 F (38-45 °C), which is lower than the melting point of beeswax, in order to facilitate rapid filtering. Honey should be strained using coarse mesh strainers at room temperature and at 110 degrees Fahrenheit, respectively.

While extracting honey, a sieve with a mesh count of six to twelve will be helpful. This honey needs to be passed through a series of strainers with 30, 50, and 80 mesh after it reaches 110 degrees Fahrenheit. Before processing honey, it is always best to indirectly heat it in a water bath, making sure the honey is kept away from the heat source from both the sides and bottom. Certain items can undergo severe discoloration, flavor loss, and decomposition if they are processed using direct heat. It has been found that 20 minutes at 145 F is sufficient to kill all of the yeast cells that ferment honey.

It is also advised to heat honey that exhibits stronger granulation tendencies to 160 degrees Fahrenheit. However, in order to maintain the quality of honey, there should be a commensurate decrease in heating time for higher processing temperatures. Processing honey should be limited to temperatures between 140 and 145 degrees Fahrenheit, as prolonged heating destroys the honey's color and aroma. Darker honeys, in particular, get much darker when cooked above 150 degrees Fahrenheit and have an overpowering molasses-like smell.

Second, longer exposure to higher temperatures causes honey to generate hydroxyl-methyl-furfural (HMF), which is what causes a positive fiehe test result. The Prevention of Food Adulteration Act prohibits the use of a positive fiehe's test. The volatile flavoring compounds in honey cause it to lose its flavor quickly when cooked in open containers. Since honey is a thick, viscous liquid, it is difficult to heat it evenly, leaving the middle of large containers of honey unheated. It is advised to heat honey in thin films and enclosed containers to preserve the honey's original flavor and color while guaranteeing that every particle is heated to the point of total destruction of yeast cells.

Following processing, honey needs to be stored in settling tanks with a strainer with an 80–110 mesh size and left to settle. In order to prevent air bubbles from entering the honey while filling the containers, the honey should be removed from the bottom of the settling tank; it should never be allowed to fall from a height. One way to achieve this is to let the honey trickle down the edges of the jar. Processed honey should be kept in storage at 60 degrees Fahrenheit. Lower temperatures cause granulation, whereas higher temperatures cause honey to lose its color and cause hydroxyl-methyl furfural (HMF) to develop.

CHAPTER 14

Recent Advances in Beekeeping

Contemporary innovations are being used by the beekeeping sector to improve apiculture and conserve the bees. These advancements in beekeeping include robotization, artificial intelligence (AI), and mobile apiary management. They make it possible to monitor swarms from beginning to end, enhance bee health, streamline hive evaluations, and guarantee the best conditions for colonies. This industry analysis explores the top trends in beekeeping and apiculture. Explore how each trend enhances bee health and quality of production by reading on.

Mobile Apiary Management

Beekeepers may maintain surveillance on several hives at once with the help of mobile apiary management. Depending on the apiculture technology being utilized, it can provide data for variables like temperature, humidity, hive weight, and even sound while on the go. This gives the apiarist the authority to address any anomalies right away in order to protect the health of the honeybee population. Additionally, transportable systems offer remote access to data from the hive. This saves time and money on physical inspections and enables farmers to manage several apiaries at once. Long-term hive management is aided by digital bee farm management solutions, which facilitate simple and effective record-keeping and offer operational insights. To help with this, up-and-coming businesses create solutions that facilitate data-sharing features for the beekeeping community. This promotes problem-solving and group learning while advancing scientific inquiry.

Apimoov Facilitates Mobile Colony Monitoring

Apiculturists can track their hives and colonies with the use of a web portal and beekeeping app offered by French startup ApiMoov. Farmers can use an interface to record observations and actions on their hives with this field recording tool. This makes it possible to plan actions for managing colonies, varroa, and queen bees as well as to track the development of the food and brood. Additionally, the system helps with artificial swarm development, royal jelly and honey yield management, and voice recognition-based offline hive monitoring. In addition, the website offers notifications for systematic digital beekeeping, an integrated summary of hives, and a study of apiary visits.

BeeTech Services Makes a Digital Beekeeping Assistant

An end-to-end apiary and management software system called BeeKing is developed by the Latvian startup Bee Tech services. It includes controlling the life cycle of colonies and queen bees, maintaining records for apiaries and hives, and making inspections easier for hive management. In order to enable quick response to changes in hive conditions and honey bee health, the software uses real-time tracking via near-field communication (NFC) tags and voice sensors. BeeKing makes managing multiple apiaries and colonies easier by offering comprehensive data analysis to support informed decision-making for increased productivity and hive health. It also offers insights into bee origin counts, varroa treatments, and the quantity of honey and pollen collected.

AI-Powered Beekeeping

Artificial intelligence enables automated processing and predictive analysis of massive amounts of data related to beekeeping. AI algorithms are used by beekeepers to predict key performance indicators like honey yields, the best times to split hives, and early detection of pests. Modern AI systems evaluate audio and visual information from hives to determine the health and activity of bees. Artificial intelligence (AI) enables beekeepers to effectively manage their resources and boost income by giving them accurate information in real-time. In this manner, beekeepers cut down on the time and labor needed for standard inspections and maintenance. Furthermore, improvements in machine learning (ML) have made it possible for researchers to comprehend bee behavior and colony health in relation to alterations in the environment and beekeeping techniques.

Humming Hive Facilitates Beehive Condition Monitoring

The hive data is then analyzed by the startup's AI and ML algorithms to find possible issues like pests, illnesses, or insufficient honey output. Moreover, when the hive atmosphere is not ideal, the startup's associated smartphone app sends out alerts. In this sense, the startup's technology supports improved hive management and honey production through bee behavior monitoring and control. A Pakistani firm called Humming Hyve is selling the Bumble Bee V1.0 beekeeping tool. This Internet of Things gadget tracks the health of beehives and the amount of honey produced. The startup's gadget has a GPS module, a load cell, and sensors for temperature and humidity. It gathers environmental data about the hive and transmits it to the cloud for processing.

Agrobee Enables AI-assisted Pollination

Agrobe, a Brazilian firm, develops a smartphone app that links farmers and apiarists to guarantee biological control and sustained pollination in crops. It uses artificial intelligence (AI) to predict when different farming cultures would pollinate most effectively, and it lets farmers request pollination from beekeepers. In the same way, beekeepers monitor pollination requests and evaluate pollinator behavior in the field using Agrobee's app. With the help of Agrobee's technology, growers can use fewer chemicals, and beekeepers can control their hives remotely, which boosts profits.

Value-added Apiculture Products

Value-added apiculture products help beekeepers diversify their revenue streams and lessen their reliance on honey production. This offers stability in terms of finances, especially when disease, pests, or bad weather cause honey output to decline. Beekeepers use the normally lost byproducts of beekeeping to create a wide range of value-added products. For instance, propolis and beeswax are used to make candles, balms, and cosmetics in health items. These products also provide substitutes for manufactured goods in a number of sectors, including food, health supplements, and cosmetics. Propolis, royal jelly, honey, and bee pollen are being used by startups more frequently in dietary supplements, medications, and gourmet meals. In this way, new businesses enable beekeepers to meet customer demand for natural products while also growing their market.

Apiceuticals Manufactures Clean Skincare Products

Clean skincare products are made possible by the exclusive PROPOWAX technology developed by the Greek startup Apiceuticals. In its beauty goods, the technology uses honeycomb as a bioactive ingredient for antioxidants. In this way, the startup ethically sources components for its range of cosmetics, adding value to bee-sourced products.

Bee Loop Makes Beeswax-Based Packaging

Bee Loop, a Lithuanian firm, offers pure beeswax honey packaging that produces no waste. The packaging container made by the firm using beeswax can be recycled and returned to Bee Loop or nearby beekeepers for further usage. It reduces the waste that packing would otherwise produce and preserves the honey in its natural state.

Precision Beekeeping

Beekeepers may exploit precision beekeeping and get real-time data on hive conditions by integrating IoT sensors, cameras, and remote monitoring. This makes it possible for beekeepers to maintain a careful eye on the weight, temperature, humidity, and other vital indicators of their hives. In order to make educated judgments, businesses also create digital beekeeping platforms that examine data trends, patterns, and correlations. Platforms for precision beekeeping enable resource allocation, hive management, and prompt responses in an emergency. Beekeepers use information from pollen availability, honey reserves, and colony strength to guide decisions about extracting honey, splitting hives, and further feeding. Beekeepers regulate hive population, prevent swarming, and sustain hive productivity by keeping an eye on hive conditions and behavior.

We'll Bee Provides AI-Powered Hive Data Processing

We'll Bee, an Italian firm, creates the Beeasy device and web platform to help the beekeeping sector. Beeasy makes use of a robotic eye that tracks individual bees as they come into or go out of a hive on its own. Its AI uses this information along with weight and internal temperature to give real-time insights about the behavior of the bees and the hive environment. We'll Bee, an Italian firm, creates the Beeasy device and web platform to help the beekeeping sector. Beeasy makes use of a robotic eye that tracks individual bees as they come into or go out of a hive on its own. Its AI uses this information along with weight and internal temperature to give real-time insights about the behavior of the bees and the hive environment.

Ecosight Technologies Advances Data-Driven Beekeeping

A multi-sensor smart hive equipment and AI-integrated platform called Beeprecise are developed by Indian startup Ecosight Technologies for precision beekeeping. Through the use of a gateway hub, the startup's smart device gathers real-time data from within and around beehives and transmits it to the cloud. Through the BeePrecise app, its AI examines this data to offer insights into hive conditions and identify potential hazards. Beeprecise, then, maximizes the beekeeping process and issues notifications regarding, among other things, infections, weak colonies, swarming, hive splitting, and pollination efficiency. Predicting honey production and other hive activities, as well as proactive hive management, are made easier with the help of this real-time data.

Bee Health Supplements

Apiculturists utilize health supplements to improve bees' general health and immune systems naturally. These concoctions provide vital vitamins, minerals, and nutrients that bees may not get in their natural habitat. By enhancing their resistance to illnesses and pests, cutting-edge nutrients help colonies grow stronger and healthier. Furthermore, by giving bees the nutrients they need, they aid in their recovery from stresses like pesticide exposure, environmental changes, or habitat loss. Ingredients in health supplements that lower tension and promote serenity in the hive are frequently used. This helps ensure the long-term survival and sustainability of the bee population. Supplements are used by apiarists, especially for brood growth and during times of high stress activities such as transportation, hive splitting, or environmental disruptions. This increases the overall strength and productivity of the hive by producing worker bees that are stronger and more productive.

Bee-Lab offers Vitamins & Supplements for Bees

In order to enhance bee health, the Greek business Bee-Lab produces vitamins and supplements. The company's Booster line of goods includes a natural bee supplement that is packed with vital micronutrients. They include vitamins and bioflavonoids that strengthen bees' immune systems. Likewise, it provides Aciflex oxalic acid stips to keep hives free of varroa mite growth and to preserve the general well-being and productivity of the colony. The natural substances used in the startup's products preserve the health of bee colonies without interfering with their regular operations.

Dalan Animal Health develops a Bee Vaccine Program

A bee vaccination program is developed by US firm Dalan Animal Health to protect queen bees from American Foulbrood (AFB). The method developed by the business exposes queen bees to dead germs through an inactive vaccination process. As a result, larvae have increased resistance to infection. The vaccination is given by combining it with queen sweets, and it has no effect on the health of the queen. The business offers tailored immunization schedules for queen cages. This helps beekeepers defend their colonies against Paenibacillus larvae, especially the newly hatched honeybee larvae.

Robotization

Another breakthrough in beekeeping that is revolutionizing the sector is roboticization, which provides precise management for beehives and automates a variety of labor-intensive chores. Beehive inspections for pest or disease identification, honey production evaluations, and colony health monitoring are handled by robots and automated systems. Robotic arms that manipulate hive frames also decrease disruption to bees and lower the possibility of hive damage. In addition to saving labor and time, automated honey extraction methods increase harvesting efficiency while preserving product quality. Robots are being utilized more and more to detect patterns and abnormalities in bee behavior in conjunction with computer vision and machine learning. This enables beekeepers to react quickly to possible problems like illnesses, parasites, or stressors in the environment. As a result, robotization maximizes the production of honey and preserves the well-being of bee colonies.

BeeWise Advances Autonomous Beehive Management

A self-contained beehive management system with an integrated smartphone app is designed by US-based startup BeeWise. With the help of AI and precise robotics, Beehome, the startup's robotic beehive, can monitor and care for each individual bee within. It also has automated agricultural harvesting, autonomous swarm prevention, insect control, and climate and humidity control. This minimizes colony loss, allows for real-time response to challenges, and enhances bee productivity by maximizing hive conditions. Additionally, the startup's automated and solar-powered hive monitoring system is installed in the fields to enhance pollination. In this way, it lowers the amount of labor and materials needed to maintain the health of honeybees and their hives.

Beeplant Makine Offers Robot Beekeeper

Robotic beekeeper and transportable hive BeePlant is created by Turkish firm Beeplant Makine. It tracks the starting and breeding colony during hatching, as well as the ages of the bee larvae, by utilizing image processing. The quantity and position of honeycomb nests that are ideal for producing royal jelly are then determined by the solution. With the use of this information, BeePlant moves the recognized bee larvae into beeswax thimbles within its mobile hive in order to extract premium royal jelly. Thus, Beeplant Makine greatly enhances harvest results while requiring less labor.

Early Pest & Disease Detection

Early pest and disease detection allows beekeepers to take swift action and stop the spread of these problems. Beekeepers can use digital technologies to diagnose pest and illness infestation problems and develop treatment plans. In addition to minimizing their effects on hive health and productivity, early detection and prompt treatments also prevent significant colony and revenue loss. Cutting-edge technology enables continuous monitoring of hive conditions, such as sensor-based monitoring or AI-powered picture recognition systems. By identifying aberrant behavior, erratic patterns, and illness indicators, these technologies enable preventative steps to preserve colony health. Additionally, businesses are creating substitute techniques, such as lasers, to identify and eliminate infestations such as varroa mites or unclean broods. By doing these steps, the usage of chemical treatments is reduced.

Combplex Provides Varroa Mite Control Tool

US-based startup Combplex develops a laser-based varroa mite infection prevention tool. The startup's product is a portable laser device that combats Varroa mites in bee colonies and alerts the beekeepers before taking preventive actions. The laser device provides a chemical-free, non-invasive alternative to otherwise chemical-based solutions. It, thus, ensures higher survival rates for bees and improves overall hive health. Beekeepers and farmers use this tool to monitor hive health remotely and manage pest infections cost-effectively.

Intelligent Hives Accelerates Hive Disease Detection

A Polish firm called Intelligent Hives creates an application for intelligent hives that permits illness diagnosis and remote hive access. Using an integrated platform, GPS locators, and Internet of Things sensors, the startup's Beehive Monitoring System keeps an eye on hive conditions in real time. Thanks to technology, diseases can be identified and beekeepers are notified of any anomalies. By effectively monitoring many hive characteristics and initiating corrective actions, it enhances bee health, boosts honey yield, and lowers labor expenses.

Sound Analysis

Sound analysis is a tool used by beekeepers to identify irregularities or behavioral shifts in their hives early on. Beekeepers can detect possible problems including queenlessness and swarming preparations by listening to the sounds made by the hive. Furthermore, beehive patterns are identified by sound analysis algorithms to aid in the early diagnosis of problems like AFB, infestations by Varroa mites, and other diseases. Prompt action and efficient management are made possible by early detection. Another non-intrusive technique that beekeepers can use to evaluate hive health and behavior without upsetting the bees is sound-based monitoring. As a result, the bees experience less stress and there are less interruptions to their regular routines. It also reduces the need for time-consuming and expensive in-person hive inspections.

OSBeehives Aids Colony Audio Analysis

A US-based firm called OSBeehives creates audio analysis tools for hive monitoring and bee inspection. The startup's acoustic sensor, called BuzzBox Mini, is designed to pick up bee colony sounds. Once the data is analyzed, its AI algorithms determine the health of the hive. Potential problems are alerted to by the device through a linked smartphone app, enabling prompt interventions. With its extensive health metrics that include inspection records, hive temperature, and humidity, it also facilitates data-driven decision-making. As a result, OSBeehives gives beekeepers electronically and remotely access to information about hive health.

AgriSound Facilitates Beehive Acoustics Monitoring

AgriSound, a UK-based startup, creates an in-hive sensor that is interference-free for sound-based hive monitoring. The business gathers essential bioacoustic data on hive changes using Internet of Things (IoT)-connected sensors. By enabling preventive actions against risks, they enable beekeepers to keep an eye on the health and productivity of their bee colonies. Additionally, the solution makes it easier to use any connected device to analyze data from the recorded bioacoustics. Furthermore, Agrisound uses Internet of Things-enabled hive scales to offer insights into honey yield forecast. The firm protects the sustainability and production of hives while advancing bee husbandry through a data-driven approach.

Remote Hive Monitoring

Beekeepers can obtain up-to-date information on temperature, humidity, hive weight, and activity levels through remote hive monitoring. This makes it possible to monitor ongoingly and to recognize any changes or anomalies right away. Cutting-edge remote monitoring technologies offer insightful data on resource management, including pollen availability and honey reserves. As a result, beekeepers are better equipped to decide whether to extract honey, prevent swarms, or provide additional nutrition. In doing so, resource allocation and hive productivity are optimized. Additionally, beekeepers save labor and time by conducting fewer in-person inspections when bee hives are monitored remotely. This reduces bee stress and is especially helpful in large-scale operations with several apiaries.

apic.ai Digitizes Bee Behaviour Monitoring

Apic.ai, a German business, provides beekeepers with an AI-based and edge computing solution for automated pollinator behavior monitoring. It uses a camera system that is not overly intrusive to track the movements of bees and bumblebees as they enter and leave the hive. The pre-trained AI in the solution examines the accumulated video material and provides insights tailored to the user's query. Solar power powers the visual camera feed, allowing for continuous monitoring even during blackouts. By evaluating pollen baskets, size, and mortality rates, it keeps track of pollinator activity within a specific time frame, feeding habits, and pollen diversity. This system automates pollinator behavioral assessments, giving beekeepers data-driven insights for product and activity design.

BeeSage Builds Smart Beehive Scales

A connected application and digital beehive scale are being offered by the Dutch startup BeeSage. To gauge production capacity, the startup's smart beehive scale is positioned beneath colonies of a moderate size. With hourly telemetry data updates, the app employs smart scale data to predict the apiary's total output capacity. The device maps position, sends web warnings, and logs beehive weight, 24-hour weight variations, and ambient temperature. This helps apiarists to plan honey collection, assess nectar flow, manage numerous colonies at once, and stop vandalism.

Sustainable Hive Design

The environmental impact of beekeeping is further minimized via hive design, even though beekeeping is already a sustainable sector. Startups concentrate on creating hives powered by renewable energy sources that offer the best possible living conditions for bees. It takes into account elements like insulation, ventilation, and space usage to establish a cozy and healthful atmosphere that supports the colony's well-being. Furthermore, elements that promote organic disease and pest management are incorporated into sustainable hive designs.

In order to minimize the need for chemical treatments, startups include screened floors for ventilation, entrance reducers, and detachable bottom boards for mite control. Incorporating queen excluders, sometimes known as swarm traps, and leaving sufficient room for colony growth are two other novel hive designs that address swarming. Additionally, the utilization of renewable and environmentally friendly materials made from recycled or sustainably sourced wood is becoming a top priority for businesses.

ValleyBee Makes Recycled Plastic Hives

ValleyBee, a Nigerian business, uses recycled plastic to create hives. The hive does not include frames, foundations, or excluders because it is made to resemble natural bee habitats. In addition, the hive has built-in sensors, a humidifier, and a heat source that allow it to automatically modify its internal temperature based on the outside weather. The firm offers a smartphone platform for remote hive monitoring and changes, in addition to colony installation and maintenance services.

Beeonic Provides Sustainable Hive Designs

Beeonic, a German business, uses natural materials to create an IoT-enabled stackable beehive. Made from sheep's wool, which naturally controls the temperature in the hive, it has a core that regulates both moisture and heat. Additionally, this lessens the possibility of mold growth and disease infection inside the beehive. The hive's sensors and cloud platform track temperature, humidity, and weight in the hive habitat. In order to maximize hive handling, this gives beekeepers a sustainable hive design that controls moisture and insulation.

Discover all Beekeeping Trends, Technologies & Startups

As climate change is impacting bee populations, startups are developing sustainable solutions for beekeeping and bee-sourced products. At the same time, 3D printing is enabling durable and lightweight hive designs. Drones and unmanned aerial vehicles (UAVs) further find application in pollinator behavior monitoring. Similarly, blockchain is empowering honey producers and extractors to establish their product authenticity and improve revenue generation. The report's list of "Apiculture Trends & Startups" barely scratches the surface of the trends we found while doing our data-driven innovation & startup scouting process. Gaining a competitive advantage in company can be achieved in large part by identifying new opportunities and developing technology to integrate into your operations.

Glossary

Abdomen - The third region of the body of a bee enclosing the honey stomach, true stomach, intestine, sting, and reproductive organs.

Absconding swarm - An entire colony of bees that abandons the hive because of disease, wax moth, excessive heat or water, lack of resources, or other reasons.

Acarine disease - The name of the disease caused by the tracheal mite (*Acarapis woodi*). See Tracheal mite.

Afterswarm - A small swarm which may leave the hive after the first or primary swarm has departed. These afterswarms usually have less bees associated with them than the primary swarm.

American foulbrood - A brood disease of honey bees caused by the spore-forming bacterium, Paenibacillus larvae. The spore stage of the bacterium can remain viable for many years, making is difficult to eliminate the disease.

Apiary - Colonies, hives, and other equipment assembled in one location for beekeeping operations; also known as a bee yard.

Apiculture - The science and art of raising honey bees.

Apis mellifera - Genus and species of the Western honey bee originating in Europe and Africa and now located around the world.

Paenibacillus larvae - The bacterium that causes American foulbrood.

Bait hive - A hive or box placed preferably in an elevated location used to attract and hopefully capture swarms.

Bee blower - A motorized blower used as one method to remove bees from honeycombs. Typically frames are not removed from supers prior to using the blower.

Bee bread - A fermented mixture of collected pollen and nectar or honey, deposited in the cells of a comb. Pollen is the primary pollen source for bees and is used especially by the nurse bees to produce royal jelly to feed the young larvae.

Bee brush - A brush or whisk broom used to gently remove bees from combs.

Bee escape - A device used to remove bees from honey supers or buildings by permitting bees to pass one way but preventing their return.

Beehive - A box or receptacle with movable frames, used for housing a colony of bees.

Bee metamorphosis - The three stages through which a bee passes before reaching maturity: egg, larva, and pupa. During the pupal stage, large fat reserves are used to transform both the internal and external anatomy of the bee.

Bee space - 3/8-inch space between combs and hive parts in which bees build no comb or deposit only a small amount of propolis. Bee spaces are used as corridors to move within the hive.

Beeswax - A complex mixture of organic compounds secreted by four pairs of special glands on the worker bee›s abdomen and used for building comb. Its melting point is from 143.6 to 147.2 degrees F.

Bee veil - A cloth or form of hat usually made of wire netting to protect the beekeeper›s head and neck from stings.

Bee venom - The poison secreted by special glands attached to the stinger of the bee.

Boardman feeder - A device for feeding bees that consists of an inverted jar with an attachment allowing access to the hive entrance.

Bottling (Honey Processing) - Gravity is used to fill containers of honey.

Bottom board - The floor of a beehive that all the other components build upon.

Brace comb - A small bit of wax built between two combs or frames to fasten them together. Brace comb is also built between a comb and adjacent wood, or between two wooden parts such as top bars.

Braula coeca - The scientific name of a wingless fly commonly known as the bee louse.

Brood - Immature bees that not yet emerged from their cells. Brood can be in the form of eggs, larvae, or pupae of different ages.

Brood chamber - The part of the hive in which the brood is reared; may include one or more hive bodies and the combs within.

Burr comb - A bit of wax built upon a comb or upon a wooden part in a hive but not connected to any other part.

Capped brood - Pupae whose cells have been sealed with a porous cover by mature bees to isolate them during their nonfeeding pupal period; also called sealed brood.

Cappings - A thin layer of wax used to cover the full cells of honey. This layer of wax is sliced from the surface of a honey-filled comb.

Castes - A term used to describe social insects of the same species and sex that differ in morphology or behavior. In honey bees there are two castes, workers and queens. The drones are a different sex and therefore not included.

Cell - The hexagonal compartment of comb built by honeybees.

Chilled brood - Bee larvae and pupae that have died from exposure to cold. This typically occurs in spring when the colony is expanding rapidly and on cold nights there aren›t enough bees to keep the brood warm.

Chunk honey - Honey cut from frames and placed in jars along with liquid honey.

Clarifying - Removing visible foreign material from honey or wax to increase its purity.

Clarifying Tank - Any tank or holding vessel that is use to temporarily store honey while the wax and other material separate from the honey.

Cluster - A large group of bees hanging together, one upon another.

Colony - All the worker bees, drones, queen, and developing brood living together in one hive or other dwelling.

Comb - A mass of six-sided cells made by honey bees in which brood is reared and honey and pollen are stored; composed of two layers united at their bases.

Comb foundation - A commercially made structure consisting of thin sheets of beeswax with the cell bases of worker cells embossed on both sides in the same manner as they are produced naturally by honey bees.

Comb honey - Honey produced and sold in the comb. It is produced either by cutting the comb from the frame or when the comb is built in special frames which allow for its easy removal.

Creamed honey - Honey which has crystallized under controlled conditions to produce a tiny crystal and a smooth texture. Often a starter or seed is used to help control the crystallization.

Crimp-wired foundation - Comb foundation which crimp wire is embedded vertically during the manufacturing of the foundation. The wire increases the strength of the foundation.

Cross-pollination - The transfer of pollen from an anther of one plant to the stigma of a different plant of the same species. Honeybees are excellent pollinators.

Crystallization - The formation of sugar crystals in honey. Syn. Granulation

Cut-comb honey - Comb honey cut into various sizes, the edges drained, and the pieces wrapped or packed individually

Dextrose - One of the two principal sugars found in honey; forms crystals during granulation. Also known as glucose.

Dividing - Splitting a colony to form two or more colonies.

Division board feeder - A wooden or plastic compartment which is hung in a hive like a frame and contains feed for bees.

Double screen - A wooden frame with two layers of wire screen to separate two colonies within the same hive, one above the other. An entrance is cut on the upper side and placed to the rear of the hive for the upper colony.

Drawn combs - Cells which have been built out by honey bees from foundation in a frame.

Drifting of bees - The failure of bees to return to their own hive in an apiary containing many colonies. Young bees tend to drift more than older bees, and bees from small colonies tend to drift into larger colonies.

Drone - The male honey bee

Drone comb - Comb measuring about four cells per linear inch that is used for drone rearing and honey storage.

Drone layer - A queen who is incapable of fertilizing eggs. As a result all brood produced is drones.

Dysentery - A condition of adult bees characterized by severe diarrhea and usually caused by starvation, low-quality food,, confinement due to poor weather conditions, or nosema infection.

European foulbrood - An infectious disease which only affects the brood of honey bees and is caused by the bacteria *Streptococcus pluton*.

Extracted honey - Honey removed from the comb.

Extraction - The process in which honey is drawn out of frames using centrifugal force with a honey extractor.

Extractor - A machine which removes honey from the cells of comb by centrifugal force.

Fermentation - The process of yeast utilizing sugar as a food, and as a byproduct, produce alcohol. Honey typically does not have enough moisture for fermentation to occur.

Fertile queen - A queen, which has been inseminated, naturally or artificially, and can lay fertilized eggs.

Field bees - Worker bees generally two to three weeks old that work to collect nectar, pollen, water, and propolis for the colony.

Filtering (Honey Harvesting) - The process of removing beeswax, propolis and even a couple bee parts from honey.

Follower board - A thin board the size of a frame that can be inserting into a hive to reduce the space available to the bees. This is done to help smaller colonies that may have trouble keeping the brood nest warm.

Frame - A piece of equipment made of either wood or plastic designed to hold the honey comb.

Fructose - The predominant simple sugar found in honey.

Fume board - A rectangular cover the size of a super which has an absorbent material on the underside. A chemical is placed on the material to drive the bees out of supers for honey removal.

Fumigilin-B - An antibiotic used in the prevention and suppression of nosema disease.

Glucose - See «Dextrose.»

Grafting - Removing a worker larva from its cell and placing it in a queen cup in order to have it reared into a queen.

Grafting tool - A needle or probe designed for transferring larvae from worker cells to a queen cells.

Granulation - The formation of sugar crystals in honey which may cause it to turn solid.

Harvest- The process of extracting honey from a hive.

Hive - The structure used by bees for a home.

Hive body - A wooden box which encloses the frames and is usually used as a brood chamber.

Hive stand - A structure that supports the hive.

Hive tool - A metal device used to open hives, pry frames apart, and scrape wax and propolis from the hive parts.

Honey - A sweet viscid material produced by bees from the nectar of flowers, composed largely of a mixture of sugars dissolved in about 17 percent water. It contains small amounts of mineral matter, vitamins, proteins, and enzymes.

Honeydew - A sweet liquid excreted by aphids, leaflhoppers, and some scale insects that is collected by bees, especially in the absence of a good source of nectar.

Honey house - Building used for extracting honey and storing equipment.

Honey stomach - A specially designed organ in the abdomen of the honey bee used for carrying nectar, honey, or water.

Increase - To add to the number of colonies, usually by dividing those on hand.

Inner cover - A lightweight cover used under a standard telescoping cover on a beehive.

Instrumental insemination - The introduction of drone spermatozoa into the genital organs of a virgin queen by means of special instruments.

Invertase - An enzyme produced by the honey bee which helps to transform sucrose to dextrose and levulose.

Larva (plural, larvae) - The second stage of bee metamorphosis; a white, legless, grublike insect.

Laying worker - A worker which lays unfertilized eggs, producing only drones, usually in colonies that are hopelessly queenless.

Levulose - See «Fructose.»

Mating flight - The flight taken by a virgin queen while she mates in the air with several drones.

Mead - Honey wine.

Migratory beekeeping - The moving of colonies of bees from one locality to another during a single season to take advantage of two or more honey flows.

Nectar - A sweet and often fragrant liquid secreted by the nectaries of plants for attracting animals. Nectar is the raw product of honey.

Nectar flow - A time when nectar is plentiful and bees produce and store surplus honey.

Nectar guide - Color marks on flowers believed to direct insects to nectar sources.

Nectaries - The glands of plants which secrete nectar, located within the flower or on other portions of the plant (extrafloral nectaries).

Nosema - A disease of the adult honey bee caused by the protozoan *Nosema apis*. The microbe destroys the gut of the bee and severe infections result in malnutrition and dysentery.

Nucleus - A hive of bees which consists of fewer frames than a typical hive and may be smaller in size. A nucleus usually consists of two to five frames of comb and used primarily for starting new colonies or rearing or storing queens; also called and commonly referred to a nuc.

Nurse bees - Young bees, three to ten days old, which feed and take care of developing brood.

Observation hive - A hive made largely of glass or clear plastic to allow for the observation of bees at work.

Package bees - A quantity of adult bees (2 to 5 pounds), with or without a queen, contained in a screened shipping cage with a food source.

PDB (Paradichlorobenzene) - Crystals used to fumigate stored combs against wax moth.

Pheromones - Chemical substances secreted from glands and used as a means of communication. Honey bees secrete many different pheromones.

Play flight - Short flight taken in front of or near the hive to acquaint young bees with their immediate surroundings.

Pollen - The male reproductive cell bodies produced by anthers of flowers. It is collected and used by honey bees as their source of protein.

Pollen basket - A flattened depression surrounded by curved hairs, located on the outer surface of a bee›s hind legs and adapted for carrying pollen to the hive.

Pollen substitute - Any material such as soybean flour, powdered skim milk, brewer›s yeast, or a mixture of these used in place of pollen as a source of protein to stimulate brood rearing. Typically feed to a hive in early spring to encourage colony expansion.

Pollen supplement - A mixture of pollen and pollen substitutes used to stimulate brood rearing typically in early spring to encourage colony expansion.

Pollen trap - A device for removing pollen loads from the pollen baskets of incoming bees.

Pollination - The transfer of pollen from the anthers to the stigrna of flowers.

Primary swarm - The first swarm to leave the parent colony, usually with the old queen (see secondary swarm).

Propolis - Sap or resinous materials collected from trees or plants by bees and used to strengthen the comb and to seal cracks; also called bee glue.

Pupa - The third stage in the development of the honey bee, during which it changes (pupates) from a larva to an adult bee.

Queen - A female bee with a fully developed reproductive system, and she is larger and longer than a worker bee.

Queen cage - A small cage in which a queen and three to five worker bees are confined for shipping and introduction into a colony.

Queen cell - A special elongated cell in which the queen is reared. It is above an inch or more long and hangs down from the comb in a vertical position.

Queen clipping - Removing a portion of one or both front wings of a queen to prevent her from flying.

Queen excluder - Metal or plastic device with spaces that permit the passage of workers but restrict the movement of drones and queens to a specific part of the hive.

Robbing - Stealing of nectar, or honey, by bees from other colonies which happens more often during a nectar dearth.

Royal jelly - A highly nutritious glandular secretion of young bees, used to feed the queen and young brood.

Sacbrood - A viral disease which affects the larva of honey bees.

Scout bees - Worker bees searching for a new source of pollen, nectar, propolis, water, or a new home for a swarm of bees.

Secondary swarm - A smaller swarm which may occur after the primary swarm has occurred.

Skep - A beehive made of twisted straw without movable frames.

Slatted rack - A wooden rack that fits between the bottom board and hive body. Bees make better use of the lower brood chamber with increased brood rearing. Congestion at the front entrance is reduced which can also reduce the swarming tendency.

Slumgum - The refuse from melted comb and cappings after the wax has been rendered or removed.

Smoker - A device in which materials are slowly burned to produce smoke (not flames) which is used to subdue bees. It is important to use a material that produces a cool smoke as not to harm the bees.

Solar wax melter - A glass-covered insulated box used to melt wax from combs and cappings by the heat of the sun.

Spur embedder - A handheld device used for embedding wires into foundation with the purpose of reinforcing the foundation.

Stinger - The modified structure of a worker honey bee used as a weapon of offense. Honey bees have a barbed stinger which stays embedded in the recipient of sting cause the bee to later die.

Streptococcus pluton - Bacteria that cause European foulbrood.

Sucrose - Principal sugar found in nectar.

Super - Any hive body, or smaller box, used for the storage of surplus honey which the beekeeper will harvest. Normally it is placed over or above the brood chamber. Betterbee offers shallow, medium, and deep supers.

Supersedure - The natural replacement of an established queen by a newly reared queen in the same hive.

Surplus honey - Honey removed from the hive which exceeds that needed by bees for their own use.

Swarm - A large number of worker bees, drones, and usually the old queen that leaves the parent colony to establish a new colony.

Swarming - The natural process of propagating a colony of honey bees.

Swarm cell - Queen cells usually found on the bottom of the combs before swarming.

Terramycin - An antibiotic used to control American and European foulbrood.

Uncapping - The process by which the thin layer of beeswax is removed, exposing honey frames for honey harvesting.

Uncapping knife - A knife used to shave or remove the cappings from combs of sealed honey prior to extraction. These can be heated by steam or electricity.

Uniting - Combining two or more colonies to form one larger colony.

Virgin queen - A queen which is not mated.

Wax glands - Glands that secrete beeswax, which are in pairs on the underside of the last four abdominal segments.

Wax moth - Larvae of the moth *Golleria mellonclia,* which can seriously damage brood and empty combs.

Winter cluster - A ball-like arrangement of adult bees within the hive during winter.

Worker bee - A female bee whose reproductive organs are undeveloped. The majority of the honey bees are worker bees and they do all the work in the colony except for laying fertile eggs.

Worker comb - Comb measuring about five cells to the inch, in which workers are reared and honey and pollen are stored.

Distributing of Bee Boxes, Accessories and Giving Training on Scientific Beekeeping in West Godavari District, AP under ICAR SCSP PROJECT

Distributing of Bee Boxes, Accessories and Giving Training on Scientific Beekeeping in West Godavari District, AP under ICAR SCSP PROJECT

Distributing of Bee Boxes, Accessories and Giving Training on Scientific Beekeeping in West Godavari District, AP under ICAR SCSP PROJECT

Distributing of Bee Boxes, Accessories and Giving Training on Scientific Beekeeping in West Godavari District, AP under ICAR SCSP PROJECT

Distributing of Bee Boxes, Accessories and Giving Training on Scientific Beekeeping in West Godavari District, AP under ICAR SCSP PROJECT

Distributing of Bee Boxes, Accessories and Giving Training on Scientific Beekeeping in West Godavari District, AP under ICAR SCSP PROJECT

Entrepreneurship Development Programme on Value Added Products of Honey and Bees Wax at Dr YSRHU - College of Horticulture VRGUDEM

Entrepreneurship Development Programme on Value Added Products of Honey and Bees Wax at Dr YSRHU - College of Horticulture VRGUDEM

Index